PELICAN BOOKS
TIME'S ARROW, TIME'S CYCLE

Stephen Jay Gould grew up in New York City. He graduated from Antioch College and received his Ph.D. from Columbia University in 1967. Since then he has been on the faculty of Harvard University. He considers himself primarily a palaeontologist and an evolutionary biologist, though he teaches geology and the history of science as well. A frequent and popular speaker on the sciences, his published work includes *Ontogeny and Phylogeny*, a scholarly study of the theory of recapitulation; *The Mismeasure of Man* (Pelican 1983), winner of the National Book Critics' Circle Award for 1982; the popular collections of essays *Ever Since Darwin: Reflections in Natural History* (Pelican 1980), which received great acclaim: 'Unreservedly, they are brilliant' – *New Scientist*; *The Panda's Thumb: More Reflections in Natural History* (Pelican 1983), which won the 1981 American Book Award for Science; *Hen's Teeth and Horse's Toes: Further Reflections in Natural History* (Pelican, 1984); *The Flamingo's Smile* (Pelican 1987); and *An Urchin in the Storm* (1988).

TIME'S ARROW

TIME'S CYCLE

Myth and Metaphor in the Discovery
of Geological Time

STEPHEN JAY GOULD

PENGUIN BOOKS

PENGUIN BOOKS

Published by the Penguin Group
27 Wrights Lane, London w8 5tz, England
Viking Penguin Inc., 40 West 23rd Street, New York, New York 10010, USA
Penguin Books Australia Ltd, Ringwood, Victoria, Australia
Penguin Books Canada Ltd, 2801 John Street, Markham, Ontario, Canada l3r 1b4
Penguin Books (NZ) Ltd, 182–190 Wairau Road, Auckland 10, New Zealand

Penguin Books Ltd, Registered Offices: Harmondsworth, Middlesex, England

First published in the USA by Harvard University Press 1987
Published in Pelican Books 1988
3 5 7 9 10 8 6 4 2

Made and printed in Great Britain by
Richard Clay Ltd, Bungay, Suffolk
Filmset in Galliard

For

RICHARD WILSON, M.D.

KAREN ANTMAN, M.D.

Sine quibus non
In the most brutal, literal sense.

Acknowledgments

The genesis of this book lies in the same conflict and interaction of metaphors—arrows of history and cycles of immanence—that fueled the discovery of deep time in geology. If I have succeeded in conveying the intended order of my thoughts, this book may strike readers as forged to some unity in a rational manner—as a product, in other words, of immanent structure reflected in the metaphor of time's cycle. But such a view, while reflecting (I hope) the logic of construction, would grossly misrepresent the psychologic of origin, for this volume is cobbled together from bits and pieces of time's arrow, quirky and unpredictable moments of my own contingent history. Tiny events that seemed meaningless at the time are horseshoe nails in the final structure. I cannot begin to specify all these incidents of "just history." My father took me to see a *Tyrannosaurus* when I was five. George White, great gentleman and bibliophile, gave me a seventeenth-century edition of Burnet's *Telluris theoria sacra* in lieu of an honorarium for a talk. John Lounsbury illustrated uniformitarianism by an example that conflated distinct meanings during an introductory course in geology at Antioch College. I knew that something was wrong, but didn't understand what, until I studied David Hume on induction. I visited the Portrush Sill in Northern Ireland on a spring field course at the University of Leeds (during an undergraduate year abroad), and saw the dichotomy of neptunism and plutonism etched in rocks. I stood in mixed horror and fascination before the skeleton(s) of Ritta-Christina, the Siamese twin girls of Sardinia, in a Paris museum. I perceived Burnet's frontispiece in the glittering beauty of

James Hampton's Throne for Christ's second coming at the National Museum of American Art. I listened to Malcolm Miller, self-appointed sage of Chartres, reading medieval metaphors in glass and statuary. R. K. Merton then showed me what a vainglorious fool I had been in thinking I had discovered the origin of Newton's phrase about the shoulders of giants in the south transept of that greatest among cathedrals.

I owe a more profound and immediate debt to colleagues who have struggled to understand the history of geology. I present this book as a logical analysis of three great documents, but it is really a collective enterprise. I am embarrassed that I cannot now sort out and properly attribute the bits and pieces forged together here. I am too close to this subject. I have taught the discovery of time for twenty years, and have read the three documents over and over again (for I regard such repetition as the best measuring stick of an intellectual life—when new insights cease, move on to something else). I simply do not remember which pieces came from my own readings of Burnet, Hutton, and Lyell, and which from Hooykaas, or Rudwick, Porter, or a host of other thinkers who have inspired me—as if exogeny and endogeny could form separate categories in any case!

In the most immediate sense, I owe great thanks to Don Patinkin, of Hebrew University, Jerusalem—and to Eitan Chernov, Danny Cohen, and Rafi Falk, guides and friends during my visit. This book is a greatly elaborated and reworked version of the first series of Harvard-Jerusalem lectures, presented at Hebrew University in April 1985. Arthur Rosenthal, director of Harvard University Press, conceived this series and brought it to fruition—and to him as godfather my deepest thanks. I can only hope that I have set a worthy beginning to a series that, by time's arrow of progress, will soon supersede its inception (while I hope for some remembrance in time's cycle of memory).

As for Jerusalem, the truly eternal city, I can only say that I finally understand Psalm 137: "If I do not remember thee, let my tongue cleave to the roof of my mouth; if I prefer not Jerusalem above my chief joy." Now that's quite a tribute from a man who lives by lecturing!

Contents

1. *The Discovery of Deep Time*   *1*

   Deep Time   1
   Myths of Deep Time   3
   On Dichotomy   8
   Time's Arrow and Time's Cycle   10
   Caveats   16

2. *Thomas Burnet's Battleground of Time*   *21*

   Burnet's Frontispiece   21
   The Burnet of Textbooks   23
   Science versus Religion?   24
   Burnet's Methodology   26
   The Physics of History   30
   Time's Arrow, Time's Cycle: Conflict and Resolution   41
   Burnet and Steno as Intellectual Partners in the Light
      of Time's Arrow and Time's Cycle   51

3. *James Hutton's Theory of the Earth: A Machine
   without a History*   *61*

   Picturing the Abyss of Time   61
   Hutton's World Machine and the Provision of Deep Time   63
   The Hutton of Legend   66
   Hutton Disproves His Legend   70
   The Sources of Necessary Cyclicity   73
   Hutton's Paradox: Or, Why the Discoverer of Deep Time
      Denied History   80
   Borges's Dilemma and Hutton's Motto   92
   Playfair: A Boswell with a Difference   93
   A Word in Conclusion and Prospect   96

4. *Charles Lyell, Historian of Time's Cycle*   99

   The Case of Professor Ichthyosaurus   99
   Charles Lyell, Self-Made in Cardboard   104
   Lyell's Rhetorical Triumph: The Miscasting
      of Catastrophism   115
   Lyell's Defense of Time's Cycle   132
   Lyell, Historian of Time's Cycle   150
   The Partial Unraveling of Lyell's World View   167
   Epilogue   178

5. *Boundaries*   181

   Hampton's Throne and Burnet's Frontispiece   181
   The Deeper Themes of Arrows and Cycles   191

   Bibliography   211

   Index   217

Illustrations

2.1 The frontispiece from the first edition of Thomas Burnet's *Telluris theoria sacra*, or *Sacred Theory of the Earth*. *20*

2.2 Burnet attempts to assess the amount of water in the oceans by the classical method of sounding. (From first edition.) *31*

2.3 Burnet's physical cause of the deluge. (From first English edition.) *33*

2.4 The earth's current surface, a product of crustal collapse during the deluge. (From first edition.) *34*

2.5 The chaos of the primeval earth as related in Genesis 1. (From first edition.) *35*

2.6 The perfect earth of the original paradise of Eden, arranged as concentric layers according to density after the descent of particles from primeval chaos. (From first edition.) *35*

2.7 The earth's surface in its paradisiacal state. (From first English edition.) *37*

2.8 The earth made perfect a second time, after the descent of particles by density into concentric layers following the future conflagration. (From first edition.) *39*

2.9 Steno's geological history of Tuscany, as rearranged in the English translation by J. G. Winter (1916). *53*

2.10 The geological history of Tuscany in Steno's original version, arranged as two parallel columns. *55*

3.1 John Clerk of Eldin's celebrated engraving of Hutton's unconformity at Jedburgh, Scotland. Courtesy of Sir John Clerk of Penicuik, from *Hutton: The Lost Drawings* (Edinburgh: Scottish Academic Press Limited). *60*

3.2 A figure from the first edition of Charles Lyell's *Principles of Geology* (1830) showing the famous locality where Hutton confirmed the igneous nature of granite by finding a multitude of granitic fingers intruding into older sediments.   *71*

4.1   De la Beche's caricature of Lyell as the future Professor Ichthyosaurus, as reproduced in the frontispiece of Frank Buckland's *Curiosities of Natural History.*   *98*

4.2   Two illustrations from the first edition of *Principles of Geology* to show Lyell's working method of comparing ancient results with modern (and visible) causes that produce the same outcomes. Top: a modern volcano in the Bay of Naples, observed in eruption during historical times. Bottom: Greek islands showing, by their topography, that they surround a volcanic vent.   *106*

4.3   Agassiz's comment of highest praise jotted alongside his criticisms in his copy of Lyell's *Principles.*   *117*

4.4   A classic example of Lyell's gradualism—the "denudation of the Weald." (From the first edition of Lyell's *Principles.*)   *121*

4.5   A modern example of destruction by erosion. The Grind of the Navir (the breach between the two sections of this sea cliff in the Shetland Islands) is widened every winter by the surge passing between. (From first edition of Lyell's *Principles.*)   *147*

4.6   Modern examples of construction by earthquakes. Top: the surface at Fra Ramondo in Calabria. Bottom: two obelisks on the facade of the convent of S. Bruno in Stefano del Bosco, Italy. (From first edition of Lyell's *Principles.*)   *148*

4.7   An illustration of the problems faced by geologists in unraveling Tertiary stratigraphy. (From first edition of Lyell's *Principles.*)   *156*

4.8   Eocene mollusk fossils used by Lyell in his statistical method for zoning the Tertiary. (Plate 3, volume III of first edition of Lyell's *Principles.*)   *162*

4.9   Miocene mollusk fossils used by Lyell to zone the Tertiary. (Plate 2, volume III of first edition of Lyell's *Principles.*)   *163*

5.1   The entire composition of James Hampton's *Throne of the Third Heaven of the Nations' Millennium General Assembly*, showing its bilateral symmetry. National Museum of American Art, Smithsonian Institution, gift of an anonymous donor.   *180*

5.2   Christ's throne, the centerpiece of Hampton's composition.   *183*

5.3   Another piece from the midline of Hampton's *Throne.*   *185*

5.4   Hampton's blackboard, showing his plan for the entire composition.   *186*

5.5   Plaques labeled B.C. and A.D., symmetrically placed about the midline of Hampton's *Throne* to illustrate the direction of time's arrow. *188–189*

5.6   Symmetrical structures from the right and left sides of Hampton's *Throne*, showing in their inscriptions the repetitions of time's cycle. *192–193*

5.7   Swainson's rigidly numerological system of taxonomy, inconceivable for an arrangement of organisms in a world of contingent history. *195*

5.8   Time's arrow of homology and time's cycle of analogy combine to produce the *Ichthyosaurus*. Courtesy Department Library Services, American Museum of Natural History. Neg. no. 313168. *199*

5.9   Ritta-Christina, the Siamese twins of Sardinia—neither two nor one person, but residing at an undefined middle of this continuum. *201*

5.10   Time's cycle in Canterbury. The tale of Lot's wife is repeated in the angel's advice to the Magi: do not return to Herod. *202*

5.11   Ceiling bosses of Norwich Cathedral. Noah in the ark corresponds with the baptism of Jesus. *203*

5.12   Painted and stained glass windows of King's College, Cambridge. Jonah emerging from the belly of the great fish corresponds with Christ rising from the tomb. Courtesy of the Provost and Fellows of King's College, Cambridge. *204–205*

5.13   From the great south window of Chartres. Time's arrow and cycle connect as the gospel writers of the New Testament are shown as dwarfs seated upon the shoulders of Old Testament prophets. *206*

5.14   From Chartres cathedral. At the end of time, the just rise to their beginning and reside in the bosom of Abraham. *207*

TIME'S ARROW, TIME'S CYCLE

Time we may comprehend, 'tis but five days elder than ourselves.

Sir Thomas Browne, *Religio Medici*, 1642.

The leading idea which is present in all our researches, and which accompanies every fresh observation, the sound to which the ear of the student of Nature seems continually echoed from every part of her works, is— Time! Time! Time!

George P. Scrope, leading British geologist, in 1827. This quote has become a virtual cliché by overuse in modern textbook epigraphs.

CHAPTER ONE

The Discovery of
Deep Time

Deep Time

Sigmund Freud remarked that each major science has made one signal contribution to the reconstruction of human thought—and that each step in this painful progress had shattered yet another facet of an original hope for our own transcendent importance in the universe:

> Humanity has in course of time had to endure from the hand of science two great outrages upon its naive self-love. The first was when it realized that our earth was not the center of the universe, but only a speck in a world-system of a magnitude hardly conceivable ... The second was when biological research robbed man of his particular privilege of having been specially created and relegated him to a descent from the animal world.

(In one of history's least modest pronouncements, Freud then stated that his own work had toppled the next, and perhaps last, pedestal of this unhappy retreat—the solace that, though evolved from a lowly ape, we at least possessed rational minds.)

But Freud omitted one of the greatest steps from his list, the bridge between spatial limitation of human dominion (the Galilean revolution), and our physical union with all "lower" creatures (the Darwinian revolution). He neglected the great temporal limitation

1

imposed by geology upon human importance—the discovery of "deep time" (in John McPhee's beautifully apt phrase). What could be more comforting, what more convenient for human domination, than the traditional concept of a young earth, ruled by human will within days of its origin. How threatening, by contrast, the notion of an almost incomprehensible immensity, with human habitation restricted to a millimicrosecond at the very end! Mark Twain captured the difficulty of finding solace in such fractional existence:

> Man has been here 32,000 years. That it took a hundred million years to prepare the world for him is proof that that is what it was done for. I suppose it is, I dunno. If the Eiffel Tower were now representing the world's age, the skin of paint on the pinnacle-knob at its summit would represent man's share of that age; and anybody would perceive that that skin was what the tower was built for. I reckon they would, I dunno.

Charles Lyell expressed the same theme in more somber tones in describing James Hutton's world without vestige of a beginning or prospect of an end. This statement thus links the two traditional heroes of deep time in geology—and also expresses the metaphorical tie of time's new depth to the breadth of space in Newton's cosmos:

> Such views of the immensity of past time, like those unfolded by the Newtonian philosophy in regard to space, were too vast to awaken ideas of sublimity unmixed with a painful sense of our incapacity to conceive a plan of such infinite extent. Worlds are seen beyond worlds immeasurably distant from each other, and beyond them all innumerable other systems are faintly traced on the confines of the visible universe.   (Lyell, 1830, 63)[1]

Deep time is so difficult to comprehend, so outside our ordinary experience, that it remains a major stumbling block to our understanding. Theories are still deemed innovative if they simply replace a false extrapolation with a proper translation of ordinary events into time's vastness. The theory of punctuated equilibrium, pro-

---

1. In all quotations I have followed modern American conventions of spelling and punctuation.

posed by Niles Eldredge and myself, is not, as so often misunderstood, a radical claim for truly sudden change, but a recognition
that ordinary processes of speciation, properly conceived as glacially
slow by the standard of our own life-span, do not translate into
geological time as long sequences of insensibly graded intermediates
(the traditional, or gradualistic, view), but as geologically "sudden"
origins at single bedding planes.

An abstract, intellectual understanding of deep time comes easily
enough—I know how many zeroes to place after the 10 when I
mean billions. Getting it into the gut is quite another matter. Deep
time is so alien that we can really only comprehend it as metaphor.
And so we do in all our pedagogy. We tout the geological mile
(with human history occupying the last few inches); or the cosmic
calendar (with *Homo sapiens* appearing but a few moments before
Auld Lang Syne). A Swedish correspondent told me that she set
her pet snail Björn (meaning bear) at the South Pole during the
Cambrian period and permits him to advance slowly toward
Malmö, thereby visualizing time as geography. John McPhee has
provided the most striking metaphor of all (in *Basin and Range*):
Consider the earth's history as the old measure of the English yard,
the distance from the king's nose to the tip of his outstretched
hand. One stroke of a nail file on his middle finger erases human
history.

How then did students of the earth make this cardinal transition
from thousands to billions? No issue can be more important in our
quest to understand the history of geological thought.

Myths of Deep Time

Parochial taxonomies are a curse of intellectual life. The acceptance
of deep time, as a consensus among scholars, spans a period from
the mid-seventeenth through the early nineteenth centuries. As
Rossi wrote (1984, ix): "Men in Hooke's times had a past of six
thousand years; those of Kant's times were conscious of a past of
millions of years." Since geology didn't exist as a separate and

recognized discipline during these crucial decades, we cannot attribute this cardinal event of intellectual history to an examination of rocks by one limited fraternity of earth scientists. Indeed, Rossi (1984) has argued persuasively that the discovery of deep time combined the insights of those we would now call theologians, archaeologists, historians, and linguists—as well as geologists. Several scholars, in this age of polymathy, worked with competence in all these areas.

In limiting my own discussion to men later appropriated by professional geologists as their own predecessors, I consciously work within the framework that I am trying to debunk (or enlarge). I am, in other words, treating the standard stories accepted by geologists for the discovery of time. Professional historians have long recognized the false and cardboard character of this self-serving mythology—and I make no claim for originality in this respect—but their message has not seeped through to working scientists, or to students.

My parochiality extends even further—to geography as well as discipline. For I have selected for intensive discussion only the three cardinal actors on the British geological stage—the primary villain and the two standard heroes.

The temporal order of these men also expresses the standard mythology about the discovery of time. Thomas Burnet, villain by taint of theological dogmatism, wrote his *Sacred Theory of the Earth* in the 1680s. The first hero, James Hutton, worked exactly a century later, writing his initial version of the *Theory of the Earth* in the 1780s. Charles Lyell, second hero and codifier of modernity, then wrote his seminal treatise, *Principles of Geology*, just fifty years later, in the 1830s. (Science, after all, does progress by acceleration, as this halving of temporal distance to truth suggests.)

The standard mythology embodies a tradition that historians dismiss with their most contemptuous label—whiggish, or the idea of history as a tale of progress, permitting us to judge past figures by their role in fostering enlightenment as we now understand it. In his *Whig Interpretation of History* (1931), Herbert Butterfield deplores the strategy of English historians allied with the Whig party who wrote the history of their nation as a progressive approach to their political ideals:

> The sin in historical composition . . . is to abstract events from their context and set them up in implied comparison with the present day, and then to pretend that by this "the facts" are being allowed to "speak for themselves." It is to imagine that history as such . . . can give us judgments of value—to assume that this ideal or that person can be proved to have been wrong by the mere lapse of time. (105–106)

Whiggish history has a particularly tenacious hold in science for an obvious reason—its consonance with the cardinal legend of science. This myth holds that science differs fundamentally from all other intellectual activity in its primary search to discover and record the facts of nature. These facts, when gathered and refined in sufficient number, lead by a sort of brute-force inductivism to grand theories that unify and explain the natural world. Science, therefore, is the ultimate tale of progress—and the motor of advance is empirical discovery.

Our geological textbooks recount the discovery of deep time in this whiggish mode, as a victory of superior observation finally freed from constraining superstition. (Each of my subsequent chapters contains a section on such "textbook cardboard," as I call it.) In the bad old days, before men rose from their armchairs to look at rocks in the field, biblical limitations of the Mosaic chronology precluded any understanding of our earth's history. Burnet represented this antiscientific irrationalism, so well illustrated by the improper inclusion of "sacred" in his titular description of our planet's history. (Never mind that he got into considerable trouble for his allegorical interpretation of the "days" of Genesis as potentially long ages.) Burnet therefore represents the entrenched opposition of church and society to the new ways of observational science.

Hutton broke through these biblical strictures because he was willing to place field observation before preconception—speak to the earth, and it shall teach thee. Two key Huttonian observations fueled the discovery of deep time—first, the recognition that granite is an igneous rock, representing a restorative force of uplift (so that the earth may cycle endlessly, rather than eroding once into ruin); and, second, the proper interpretation of unconformities as boundaries between cycles of uplift and erosion (providing direct evidence for episodic renewal rather than short and unilinear decrepitude).

But the world was not ready for Hutton (and he was too lousy a writer to persuade anyone anyway). Thus, the codification of deep time awaited the great textbook of Charles Lyell, *Principles of Geology* (1830–1833). Lyell triumphed by his magisterial compendium of factual information about rates and modes of current geological processes—proving that the slow and steady operation of ordinary causes could, when extended through deep time, produce all geological events (from the Grand Canyon to mass extinctions). Students of the earth could now reject the miraculous agents that compression into biblical chronology had required. The discovery of deep time, in this version, becomes one of history's greatest triumphs of observation and objectivity over preconception and irrationalism.

Like so many tales in the heroic mode, this account of deep time is about equally long on inspiration and short on accuracy. Twenty-five years after N. R. Hanson, T. S. Kuhn, and so many other historians and philosophers began to map out the intricate interpenetrations of fact and theory, and of science and society, the rationale for such a simplistic one-way flow from observation to theory has become entirely bankrupt. Science may differ from other intellectual activity in its focus upon the construction and operation of natural objects. But scientists are not robotic inducing machines that infer structures of explanation only from regularities observed in natural phenomena (assuming, as I doubt, that such a style of reasoning could ever achieve success in principle). Scientists are human beings, immersed in culture, and struggling with all the curious tools of inference that mind permits—from metaphor and analogy to all the flights of fruitful imagination that C. S. Peirce called "abduction." Prevailing culture is not always the enemy identified by whiggish history—in this case the theological restrictions on time that led early geologists to miracle-mongering in the catastrophist mode. Culture can potentiate as well as constrain—as in Darwin's translation of Adam Smith's laissez-faire economic models into biology as the theory of natural selection (Schweber, 1977). In any case, objective minds do not exist outside culture, so we must make the best of our ineluctable embedding.

It is important that we, as working scientists, combat these myths of our profession as something superior and apart. The myths may serve us well in the short and narrow as rationale for a lobbying strategy—give us the funding and leave us alone, for we know what we're doing and you don't understand anyway. But science can only be harmed in the long run by its self-proclaimed separation as a priesthood guarding a sacred rite called *the* scientific method. Science is accessible to all thinking people because it applies universal tools of intellect to its distinctive material. The understanding of science—one need hardly repeat the litany—becomes ever more crucial in a world of biotechnology, computers, and bombs.

I know no better way to illustrate this ecumenicism of creative thought than the debunking (in a positive mode) of remaining cardboard myths about science as pure observation and applied logic, divorced from realities of human creativity and social context. The geological myth surrounding the discovery of deep time may be the most persistent of remaining legends.

This book respects the defined boundaries of the myth to disperse it from within. I analyze in detail the major texts of three leading actors (one villain, two heroes), trying to find a key that will unlock the essential visions of these men—visions lost by a tradition that paints them as enemies or avatars of progress in observation. I find this key in a dichotomy of metaphors that express conflicting views about the nature of time. Burnet, Hutton, and Lyell all struggled with these ancient metaphors, juggling and juxtaposing until they reached distinctive views about the nature of time and change. These visions fueled the discovery of deep time as surely as any observation of rocks and outcrops. The interplay of internal and external sources—of theory informed by metaphor and observation constrained by theory—marks any major movement in science. We can grasp the discovery of deep time when we recognize the metaphors underlying several centuries of debate as a common heritage of all people who have ever struggled with such basic riddles as direction and immanence.

On Dichotomy

Any scholar immersed in the details of an intricate problem will tell you that its richness cannot be abstracted as a dichotomy, a conflict between two opposing interpretations. Yet, for reasons that I do not begin to understand, the human mind loves to dichotomize—at least in our culture, but probably more generally, as structuralist analyses of non-Western systems have demonstrated. We can extend our own tradition at least to the famous aphorism of Diogenes Laertius: "Protagoras asserted that there were two sides to every question, exactly opposite to each other."

I used to rail against these simplifications, but now feel that another strategy for pluralism might be more successful. I despair of persuading people to drop the familiar and comforting tactic of dichotomy. Perhaps, instead, we might expand the framework of debates by seeking other dichotomies more appropriate than, or simply different from, the conventional divisions. All dichotomies are simplifications, but the rendition of a conflict along differing axes of several orthogonal dichotomies might provide an amplitude of proper intellectual space without forcing us to forgo our most comforting tool of thought.

The problem is not so much that we are driven to dichotomy, but that we impose incorrect or misleading divisions by two upon the world's complexity. The inadequacy of some dichotomies rests upon their anachronism. Darwin, for example, built such a prominent watershed that we tend to impose the conventional dichotomy of his achievement—evolution versus creation—backward into time, forcing it upon different debates about other vital subjects. Examples are legion, and I have treated several in my essays—from a rampant precursoritis that tries to find Darwinian seeds in Greek thought; to the search for evolutionary tidbits in pre-Darwinian works, leading us to ignore, for example, an extensive and subtle treatise about embryology for a fleeting passage about change (see Gould, 1985, on Maupertuis); to the miscasting as creationist of a great tradition in structural biology (from Geoffroy Saint-Hilaire to Richard Owen) because its theory of *change* denied an environmental underpinning and therefore seemed antievolutionary to some who equated transmutation itself with later views about its mechanisms (Gould, 1986b, on Richard Owen).

Other misleading dichotomies are mired in the tradition of whiggish history in science, including the divisions that have so badly miscast the history of geology and its discovery of deep time: uniformitarianism/catastrophism, empiricist/speculator, reason/revelation, true/false. Lyell, as we shall see, established much of the rhetoric for these divisions, but we have been led astray by following him uncritically.

I do not wish to argue that other dichotomies are "truer." Dichotomies are useful or misleading, not true or false. They are simplifying models for organizing thought, not ways of the world. Yet I believe, for reasons I shall outline in the next section, that one neglected dichotomy about the nature of time has particular value in unlocking the visions of my three key actors in the drama of deep time.

All great theories are expansive, and all notions so rich in scope and implication are underpinned by visions about the nature of things. You may call these visions "philosophy," or "metaphor," or "organizing principle," but one thing they are surely not—they are not simple inductions from observed facts of the natural world. I shall try to show that Hutton and Lyell, traditional discoverers of deep time in the British tradition, were motivated as much (or more) by such a vision about time, as by superior knowledge of rocks in the field. Indeed, I shall show that their visions stand prior—logically, psychologically, and in the ontogeny of their thoughts—to their attempts at empirical support. I shall also show that Thomas Burnet, villain of Whig history, tried to balance the two poles of a dichotomy that Hutton and Lyell read as victory for one side—and that, in many ways, Burnet's reading commends itself more to our current attention. Deep time, in other words, imposed a vision of reality rooted in ancient traditions of Western thought, as much as it reflected a new understanding of rocks, fossils, and strata.

This crucial dichotomy embodies the deepest and oldest themes in Western thought about the central subject of time: linear and circular visions, or time's arrow and time's cycle.

Time's Arrow and Time's Cycle

We live embedded in the passage of time—a matrix marked by all possible standards of judgment: by immanent things that do not appear to change; by cosmic recurrences of days and seasons; by unique events of battles and natural disasters; by an apparent directionality of life from birth and growth to decrepitude, death, and decay. Amidst this buzzing complexity, interpreted by different cultures in so many various ways, Judeo-Christian traditions have struggled to understand time by juggling and balancing two ends of a primary dichotomy about the nature of history. In our traditions, these poles have received our necessary attention because each captures an unavoidable theme in the logic and psychology of how we understand history—the twin requirements of uniqueness to mark moments of time as distinctive, and lawfulness to establish a basis for intelligibility.

At one end of the dichotomy—I shall call it time's arrow—history is an irreversible sequence of unrepeatable events. Each moment occupies its own distinct position in a temporal series, and all moments, considered in proper sequence, tell a story of linked events moving in a direction.

At the other end—I shall call it time's cycle—events have no meaning as distinct episodes with causal impact upon a contingent history. Fundamental states are immanent in time, always present and never changing. Apparent motions are parts of repeating cycles, and differences of the past will be realities of the future. Time has no direction.

I present nothing original here. This contrast has been drawn so often, and by so many fine scholars, that it has become (by the genuine insight it provides) a virtual cliché of intellectual life. It is also traditional—and central to this book as well—to point out that Judeo-Christian traditions have struggled to embrace the necessary parts of both contradictory poles, and that time's arrow and time's cycle are both prominently featured in the Bible.

Time's arrow is the primary metaphor of biblical history. God creates the earth once, instructs Noah to ride out a unique flood in a singular ark, transmits the commandments to Moses at a distinctive moment, and sends His son to a particular place at a definite time to die for us on the cross and rise again on the third day. Many scholars have identified time's arrow as the most important and distinctive contribution of Jewish thought, for most other systems, both before and after, have favored the immanence of time's cycle over the chain of linear history.

But the Bible also features an undercurrent of time's cycle, particularly in the book of Ecclesiastes, where solar and hydrological cycles are invoked in metaphor to illustrate both the immanence of nature's state ("there is no new thing under the sun"), and the emptiness of wealth and power, for riches can only degrade in a world of recurrence—vanity of vanities, saith the Preacher.

The sun also ariseth, and the sun goeth down, and hasteth to his place where he arose. The wind goeth toward the south, and turneth about unto the north; it whirleth about continually, and the wind returneth again according to his circuits. All the rivers

run into the sea; yet the sea is not full; unto the place from
whence the rivers come, thither they return again . . . The thing
that hath been, it is that which shall be; and that which is done
is that which shall be done . . .   (Ecclesiastes 1:5-9)

Although both views coexist in this primary document of our
culture, we can scarcely doubt that time's arrow is the familiar or
"standard" view of most educated Westerners today. This metaphor
dominates the Bible itself, and has only increased in strength since
then—gaining a special boost from ideas of progress that have
attended our scientific and technological revolutions from the seventeenth century onward. Richard Morris writes in his recent study
of time:

Ancient peoples believed that time was cyclic in character . . .
We, on the other hand, habitually think of time as something
that stretches in a straight line into the past and future . . . The
linear concept of time has had profound effects on Western
thought. Without it, it would be difficult to conceive of the idea
of progress, or to speak of cosmic or biological evolution.
(1984, 11)

When I argue that time's arrow is our usual view, and when I
designate the idea of distinctive moments in irreversible sequence
as a prerequisite for intelligibility itself (see p. 80), please note that
I am discussing a vision of the nature of things bound by both
culture and time. As Mircea Eliade argues in the greatest modern
work on arrows and cycles, *The Myth of the Eternal Return* (1954),[2]
most people throughout history have held fast to time's cycle, and
have viewed time's arrow as either unintelligible or a source of
deepest fear. (Eliade titles his last section "the terror of history.")
Most cultures have recoiled from a notion that history embodies
no permanent stability and that men (by their actions of war), or

2. In his subtitle, *Cosmos and History*, Eliade contrasts his one-word epitomes of
the visions of time's cycle and time's arrow.

natural events (by their consequences of fire and famine) might be reflecting the essence of time—and not an irregularity subject to repeal or placation by prayer and ritual. Time's arrow is the particular product of one culture, now spread throughout the world, and especially "successful," at least in numerical and material terms. "Interest in the 'irreversible' and the 'new' in history is a recent discovery in the life of humanity. On the contrary, archaic humanity . . . defended itself, to the utmost of its powers, against all the novelty and irreversibility which history entails" (Eliade, 1954, 48).

I recognize as well that time's arrow and time's cycle are not only culture-bound but also oversimplified as catch-alls for complex and varied attitudes. In particular, Eliade shows that each pole of this dichotomy conflates at least two different versions, related in essence to be sure, but with important distinctions. Time's cycle may refer to true and unchanging permanence or immanent structure (Eliade's "archetype and repetition"), or to recurring cycles of separable events precisely repeated. Similarly, the ancient Hebrew view of time's arrow as a string of unique events between two fixed points of creation and termination is quite different from the much later notion of inherent direction (usually a concept of universal progress, but sometimes a one-way path to destruction, as in the earth's thermodynamic "heat-death" expected by catastrophists of Lyell's day as a consequence of continual cooling from an originally molten state). Uniqueness and direction are both folded into our modern idea of time's arrow, but they arose at different times and in disparate contexts.

The contrast of arrows and cycles lies so deep in Western thinking about time that a movement as central as the discovery of geological time could scarcely proceed uninfluenced by these ancient and persistent visions. I shall try to show that metaphors of time's arrow and time's cycle formed a focus for debate, and proved as fundamental to the formulation of deep time as any observation about the natural world. If we must have dichotomies, time's arrow and time's cycle is "right"—or at least maximally useful—as a framework for understanding geology's greatest contribution to human thought. I do not make this claim *a priori* or on principle, but for four specific reasons that I defend throughout this work.

First, time's arrow and time's cycle may be too simple and too limited, but it was at least *their* dichotomy—the context recognized by Burnet, Hutton, and Lyell themselves, rather than an anachronistic or moralistic contrast imposed by whiggish histories of textbook cardboard (observation/speculation, or uniformity/catastrophe).

Second, we have lost this well-articulated context of their own understanding because the pole of time's cycle has become so unfamiliar today that we no longer recognize its guiding sway upon our heroes (especially when we view them merely as superior observers with an essentially modern cast of mind). Moreover, time's cycle embodies basic principles of interpretation that we need to recover (or at least not dismiss as empirically inadequate). Eliade, great student of myth, lauded the reintroduction of time's cycle into some modern theories not because he could judge their truth but because he so well understood the deeper meaning of this metaphor:

> The reappearance of cyclical theories in contemporary thought is pregnant with meaning. Incompetent as we are to pass judgment upon their validity, we shall confine ourselves to observing that the formulation, in modern terms, of an archaic myth betrays at least the desire to find a meaning and a transhistorical justification for historical events. (1954, 147)

Third, I became convinced about the fundamental character of this dichotomy because it unlocked (at least for me) the central meaning of three great documents that I had read many times but never understood in a unified way. Parts that I had viewed as disparate fell into a piece; I was able to reorder false alignments designated by the whiggish dichotomies, and to read these texts with a better taxonomy that expressed the authors' own visions.

The test of any organizing principle is its success in rendering specifics, not its status as abstract generality. Time's arrow and time's cycle unlocked particulars of each text, and permitted me to grasp the central character of themes usually cast aside as peripheral, or not acknowledged at all.

For Burnet, I could apprehend his text (and his frontispiece) as a battleground of internal struggle and uneasy union between the two metaphors. I could understand the deeper unity between his view of the earth and Steno's in the *Prodromus*—though these two texts are usually read as opposite poles of the inappropriate archaic/modern dichotomy. For Hutton, I finally grasped his vision as time's cycle in its purest form, and I discovered a key difference between him and his Boswell, John Playfair—a distinction centered on the dichotomy of arrows and cycles, but previously invisible without this context. For Lyell, I understood the deeper themes behind his method for dating Tertiary rocks, and knew at last why he had made a mere technique the centerpiece of a theoretical treatise. And I grasped the reason for his later allegiance to evolution—as a conservative strategy of minimal retreat from his vision of time's cycle, not as the testimony of a deputy in Darwin's radical crusade.

In a larger sense, time's arrow and time's cycle became the central focus of this book when I recognized that Hutton's and Lyell's preference for deep time arose, first and foremost, from their commitment to the unfamiliar view of time's cycle, and not (as the myth professes) from superior knowledge of rocks in the field. We, in our world of time's arrow, will never understand the twin "fathers" of our profession unless we recover their vision and their metaphor.

Fourth, time's arrow and time's cycle is, if you will, a "great" dichotomy because each of its poles captures, by its essence, a theme so central to intellectual (and practical) life that Western people who hope to understand history must wrestle intimately with both—for time's arrow is the intelligibility of distinct and irreversible events, while time's cycle is the intelligibility of timeless order and lawlike structure. We must have both.

Caveats

This book has a limited, and rather self-serving, domain and purpose. It is no conventional work of scholarship, but a quest for personal understanding of key documents usually misinterpreted (at least by me in my first readings, before I grasped the role of vision and metaphor in science). I claim absolutely no originality for the theme of time's arrow and time's cycle—for this dichotomy has been explored by many students of time, from Mircea Eliade, Paolo Rossi, J. T. Fraser, and Richard Morris in our generation, back through Nietzsche to Plato. Many historians of geology (from Reijer Hooykaas, to C. C. Gillispie, to M. J. S. Rudwick, to G. L. Davies and others) have also recognized its influence, but have not worked out its full sway through textual analysis.

In addition, this book uses an almost reactionary method that, I pray, will not offend my colleagues in the history of science. It rests, first of all, as Rossi (1984) exemplifies so well in contrast, on restrictive taxonomies. The discovery of time is scarcely the work of three thinkers in Great Britain (and I use them only because I am trying to disperse the traditional myth from within). Moreover, I have followed the limited and unfashionable method of *explication des textes*. This work is a close analysis of the central logic in the first editions of three seminal documents in the history of geology. I do not maintain that such a myopic procedure can substitute for true history, especially since the greatest contemporary advances in our understanding of science have emerged from the opposite strategy of expansive analysis and exploration of social contexts. My admiration for this work is profound. I could not have begun to conceive this book without insights provided by broadened horizons that this expansive work has provided for all of us. I do understand that Burnet's rationale cannot be comprehended without the context of England in the Glorious Revolution (which intervened between the publication of his treatises on the earth's past and future), and especially his battles with radical millenarians of his day. I also appreciate that to discuss James Hutton without the Edinburgh of David Hume, Adam Smith, and James Watt is (to cite the tale of another Scotsman) to rip a child untimely from its mother's womb.

Still, I see some value in the venerable method of *explication*. The social and psychological sources of a text are manifold—the reasons why it exists at all, and why it espouses one view of the world rather than another. But truly great works also have an internal logic that invites analysis in its own terms—as a coherent argument contained within itself by brilliance of vision and synthesis of careful construction. All the pieces fit once you grasp this central logic.

I would go further and argue that the salutary theme of social context has sometimes driven us away from the logic of documents, for we decompose parts into disparate aspects of a broader setting and sometimes forget that they also cohere *inside* the work in an almost organic way—as if the covers of a book work like the skin of an organism. (We must strive to understand any creature's ecology from the outside, but morphologists, from Goethe to Geoffroy to Owen to D'Arcy Thompson, have also understood the value of structural analysis from within.) Great arguments have a universality (and a beauty) that transcends time—and we must not lose internal coherence as we strive to understand the social and psychological whys and wherefores.

I don't think that we can be accused of unreconstructed whiggery if we seek guidance and modern understanding from great arguments of the past—for examples of true wisdom are few and far between, and we need all the cases we can get. Also, as I argue above, the discovery of time was so central, so sweet, and so provocative, that we cannot hope to match its import again. The texts of this discovery will remain our most precious and instructive documents because they embody a breadth of vision and passion that cannot recur in quite the same way. Finally, something so basic and fundamental that we often neglect to say it: the study of major texts by great thinkers needs no rationale beyond the pure pleasure that such intellectual power provides. The main motivation for my strategy was simple joy.

Although my basic procedure might seem restrictive, I have tried to provide expansion by several subthemes. In particular, if texts are unified by a central logic of argument, then their pictorial illustrations are integral to the ensemble, not pretty little trifles included only for aesthetic or commercial value. Primates are visual animals, and (particularly in science) illustration has a language and set of conventions all its own. Rudwick (1976), in his most brilliant article, has developed this theme, but scholars have been slow to add another dimension to their traditional focus upon words alone. Within my theme of metaphor and vision brought to a world of observation, pictorial summary assumes an especially vital role. I found that pictures provided a key to my understanding of time's arrow and time's cycle as the primary field of intellectual struggle. When I apprehended the complexity of Burnet's frontispiece, I possessed the outline of this book. I shall therefore begin each substantive chapter by discussing a crucial picture—usually misunderstood or ignored—that captures the metaphor of time favored by each protagonist.

When Goethe, as an old man, attended the greatest debate of another dichotomy in 1830, he recognized that the arguments in l'Académie des sciences might be more important in the long run than the political revolution then engulfing the streets of Paris. For Cuvier and Geoffroy, France's greatest biologists, were thrashing out the central dichotomy of structural versus functional approaches to form (not fighting the nascent conflict of evolution versus creation, as a subsequent tradition of anachronism would assert). Goethe understood from the core of his own practice that art and science could be adjacent facets of one intellectual ensemble; he knew the passion of science as a struggle of ideas, not only a compendium of information. Goethe also realized that some dichotomies must interpenetrate, and not struggle to the death of one side, because each of their opposite poles captures an essential property of any intelligible world. He wrote of structural and functional biology (though we might as well read time's arrow and time's cycle): "The more vitally these two functions of the mind are related, like inhaling and exhaling, the better will be the outlook for the sciences and their friends."

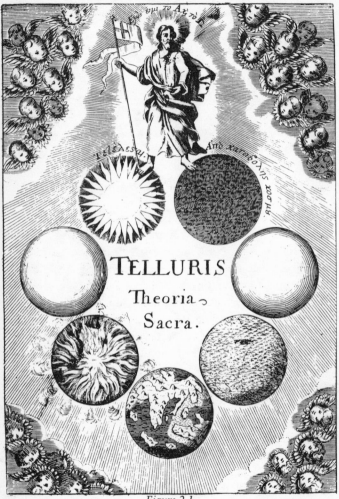

Figure 2.1
The frontispiece from the first edition of Thomas Burnet's *Telluris theoria sacra*, or *Sacred Theory of the Earth*.

CHAPTER TWO

Thomas Burnet's
Battleground of Time

Burnet's Frontispiece

The frontispiece to Thomas Burnet's *Telluris theoria sacra (The Sacred Theory of the Earth)* may be the most comprehensive and accurate epitome ever presented in pictorial form—for it presents both the content of Burnet's narrative and his own internal debate about the nature of time and history (Figure 2.1).

Below the requisite border of cherubim (for Burnet's baroque century), we see Jesus, standing atop a circle of globes, his left foot on the beginning, his right on the culmination of our planet's history. Above his head stands the famous statement from the Book of Revelation: I am alpha and omega (the beginning and the end, the first and the last). Following conventions of the watchmakers' guild, and of eschatology (with bad old days before salvation to the left, or sinister, side of divinity), history moves clockwise from midnight to high noon.

We see first (under Christ's left foot) the original chaotic earth "without form and void," a jumble of particles and darkness upon the face of the deep. Next, following the resolution of chaos into a series of smooth concentric layers, we note the perfect earth of Eden's original paradise, a smooth featureless globe. But the deluge arrives just in time to punish our sins, and the earth is next consumed by a great flood (yes, the little figure just above center is

21

Noah's ark upon the waves). The waters retreat, leaving the cracked crust of our current earth, "a broken and confused heap of bodies." In times to come, as the prophets foretold, the earth shall be consumed by fire, then made smooth again as descending soot and ashes reestablish concentric perfection. Christ shall reign for a thousand years with his resurrected saints on this new globe. Finally, after a last triumphant battle against evil forces, the final judgment shall allocate all bodies to their proper places, the just shall ascend to heaven, and the earth (under Christ's right foot), no longer needed as a human abode, shall become a star.

This tale embodies time's arrow at its grandest—a comprehensive rip-roaring narrative, a distinctive sequence of stages with a definite beginning, a clear trajectory, and a particular end. Who could ask for a better story?

But Burnet's frontispiece records more than time's arrow. The globes are arranged as a circle, not a line or some other appropriate metaphor of exclusively sequential narrative—and Christ, the Word who was with God at the beginning, straddles the inception and culmination. Consider also, the careful positioning of globes, with our current earth in the center between two symmetrical flanks. Note the conscious correspondences between right and left flanks: the perfect earth following descent of the elements from chaos (at 3:00), and directly across at 9:00, the earth made perfect again after particles descend from the conflagration; or the earth *in extremis,* first by water then by fire, on either side of its current ruined state.

In other words, Burnet displays his narrative (time's arrow) in the context of time's cycle—an eternal divine presence at top, a circular arrangement of globes beginning and ending in immanence, a complex set of correspondences between our past and future.

This same picture also embodies, and equally well, the dubious reasons for Burnet's status as a primal villain of the history of geology, a symbol of the major impediment to its discovery of deep time. For we see the earth's history intimately entwined with, indeed dictated by, a strictly literal reading of the sacred text.

The Burnet of Textbooks

Burnet emerges from our textbooks as the archetype of a biblical idolatry that reined the progress of science. We may extend this tradition of commentary right back to the other two protagonists of this book—to James Hutton, who wrote of Burnet, "This surely cannot be considered in any other light than as a dream, formed upon a poetic fiction of a golden age" (1795, I, 271); and to Charles Lyell, who remarked that "even Milton had scarcely ventured in his poem to indulge his imagination so freely . . . as this writer, who set forth pretensions to profound philosophy" (1830, 37).

No one professed the empiricist faith in purer form than the leading Scottish geologist, Archibald Geikie. His *Founders of Geology* (1897) promoted the tradition of heroes as field workers, and villains as speculators. As a "standard" history of geology for several generations, this book became the source for much continuing textbook dogma. Geikie included Burnet's book among the "monstrous doctrines" that infested late-seventeenth-century science: "Nowhere did speculation run so completely riot as in England with regard to theories of the origin and structure of our globe" (1905 ed., 66). Geikie then presented his empiricist solution—that facts must precede theory—to this retrospective dilemma: "It was a long time before men came to understand that any true theory of the earth must rest upon evidence furnished by the globe itself, and that no such theory could properly be framed until a large body of evidence had been gathered together" (1905 ed., 66).

Horace B. Woodward, in his official history of the Geological Society of London (1911, 13) placed Burnet's work among the "romantic and unprofitable labors" of its time. From a peculiar source came the most interesting of all critiques. George McCready Price, grandfather and originator of the pseudoscience known to its adherents by the oxymoron "scientific creationism," considered Burnet a special threat to his system. Price wished to affirm biblical literalism by an inductive approach based strictly on fieldwork. On the old principle that the enemy within is more dangerous than the enemy without, Price wanted to distance himself as far as possible from men like Burnet, who told their scriptural history of the earth from their armchairs:

> Their wild fancies deserve to be called travesties alike on the Bible and on true science; and the word "diluvial" has been a term to mock at ever since. Happy would it have been for the subsequent history of all the sciences, if the students of the rocks had all been willing patiently to investigate the records, and hold their fancies sternly in leash until they had gathered sufficient facts upon which to found a true induction or generalization.   (Price, 1923, 589)

This characterization persists into our generation. Fenton and Fenton, in their popular work *Giants of Geology* (1952, 22), dismiss Burnet's theory as "a series of queer ideas about earth's development," and misread his mechanism as a series of divine interventions: "Thomas Burnet thought an angry God had used the sun's rays as a chisel to split open the crust and let the central waters burst forth upon an unrepentant mankind." Davies (1969, 86), in his excellent history of British geomorphology, states that the scriptural geologies of Burnet and others "have always had a peculiar fascination for historians as bizarre freaks of pseudo-science."

Science versus Religion?

The matrix that supports this canonical mischaracterization of Burnet is the supposed conflict, or war, between science and religion. Though scholars have argued *ad nauseam* that no such dichotomy existed—that the debate, if it expressed any primary division, separated traditionalists (mostly from the church) and modernists (including most scientists, but always many churchmen as well)—this appealing and simplistic notion persists.

The *locus classicus* of "the warfare of science with theology" is the two-volume work (1896) of the same name by Andrew Dickson

White, president of Cornell University. White, although personally devout, held an even stronger commitment to the first amendment and sought to establish a nondenominational university. Speaking of his work with Ezra Cornell, he wrote: "Far from wishing to injure Christianity, we both hoped to promote it; but we did not confound religion with sectarianism" (1896, vii). White then presented his central thesis as a paragraph in bold italics:

> In all modern history, interference with science in the supposed interest of religion, no matter how conscientious such interference may have been, has resulted in the direst evils both to religion and to science, and invariably; and, on the other hand, all untrammelled scientific investigation, no matter how dangerous to religion some of its stages may have seemed for the time to be, has invariably resulted in the highest good both of religion and of science.   (1896, viii)

White began his book with a metaphor. As a member of the United States embassy in Russia, he watches from his room above the Neva in St. Petersburg as a crowd of Russian peasants breaks the ice barrier that still dams the river as the April thaws approach. The peasants are constructing hundreds of small channels through the ice, so that the pent-up river may discharge gradually and not vent its fury in a great flood caused by sudden collapse of the entire barrier:

> The waters from thousands of swollen streamlets above are pressing behind [the ice dam]; wreckage and refuse are piling up against it; every one knows that it must yield. But there is a danger that it may . . . break suddenly, wrenching even the granite quays from their foundations, bringing desolation to a vast population . . . The patient mujiks are doing the right thing. The barrier, exposed more and more to the warmth of spring by the scores of channels they are making, will break away gradually, and the river will flow on beneficent and beautiful.

The rising waters, White tells us, represent "the flood of increased knowledge and new thought"; the dam is dogmatic religion and unyielding convention (White then confesses the hope that his book might act as a mujik's channel to let light through gently). For if dogma stands fast, and the dam breaks (as truth cannot be forestalled forever), then the flood of goodness, by its volume alone, will overwhelm more than darkness: ". . . a sudden breaking away, distressing and calamitous, sweeping before it not only outworn creeds and noxious dogmas, but cherished principles and ideals, and even wrenching out most precious religious and moral foundations of the whole social and political fabric" (1896, vi).

Burnet, in White's view, was part of the dam—an example of religion's improper intrusion into scientific matters and, therefore, a danger to gentle enlightenment. This interpretation underlies the short-takes of our textbooks and classroom lectures. Modern scholars know better, but the world of textbooks is a closed club, passing its errors directly from generation to generation.

Burnet's Methodology

The Reverend Thomas Burnet was a prominent Anglican clergyman who became the private chaplain of King William III. Between 1680 and 1690, Burnet published, first in Latin then in English, the four books of *Telluris theoria sacra,* or *The Sacred Theory of the Earth: Containing an Account of the Original of the Earth, and of all the General Changes which it hath already undergone, or is to undergo Till the Consummation of all Things.* In Book I on the deluge, Book II on the preceding paradise, Book III on the forthcoming "burning of the world," and Book IV "concerning the new heavens and new earth," or paradise regained after the conflagration, Burnet told our planet's story as proclaimed by the unfailing concordance of God's words (the sacred texts) and his works (the objects of nature).

In previous hints of my affection for Burnet, I hope I did not convey the impression that I would defend him as a scientist under contemporary standards invoked by his textbook critics. In these terms, he clearly fails just as his detractors insist. *The Sacred Theory of the Earth* contains precious few appeals to empirical information. It speaks with as much confidence, and at comparable length, about an unobservable future as about a confirmable past. Its arguments cite scripture as comfortably and as often as nature. But how can we criticize Burnet for mixing science and religion when the taxonomy of his times recognized no such division and didn't even possess a word for what we now call science? Burnet, who won Newton's high praise for his treatise, was an exemplary representative of a scholarly style valued in his own day. To be sure, that style imposed stringent limits upon what we would now call empirical truth, but retrospective history, with its anachronistic standards, can only lead us to devalue (and thereby misunderstand) our predecessors—for time's arrow asserts its sway upon human history primarily through the bias of progress and leads us to view the past as ever more inadequate the further back we go.

I propose to treat Burnet with elementary respect, to take the logic of his argument seriously and at face value.[1] Burnet proceeded by a method used in our era only by Immanuel Velikovsky (among names well known). Velikovsky began his radical, and now disproven, reconstruction of cosmology and human history with a central premise that reversed our current tradition of argument: suppose, for the sake of investigation, that everything in the written documents of ancient civilizations is true. Can we then invent a physics that would yield such results?[2] (If Joshua said that the sun

1. I know that motives are ever so much more complex than the logic of argument. I accept many of the arguments advanced by scholars to untangle the hidden agenda of Burnet's conclusions—that, for example, his insistence upon resurrection only *after* a future conflagration served as a weapon against religious radicals who preached an imminent end to the world. Yet I find personal merit in taking unfamiliar past arguments at face value and working through their logic and implications. These exercises have taught me more about thinking in general than any explicit treatise on principles of reasoning.

2. While all scientists now argue that the possibilities of physics set prior limits upon what claims of the ancient texts might be historically true.

stood still upon Gibeon, then something stopped the earth's rotation—the close passage of wandering Mars or Venus in Velikovsky's reconstruction.)

Burnet began by assuming that only one document—the Bible—is unerringly true.[3] His treatise then becomes a search for a physics of natural causes to render these certain results of history. (Burnet, of course, differs from Velikovsky in a fundamental way. Velikovsky took the veracity of ancient texts only as a heuristic beginning. For Burnet, the necessary concordance of God's words and works established harmony between physics and scripture as necessary *a priori*.)

Within this constraint of concordance, Burnet followed a strategy that placed him among the rationalists ("good guys" for the future development of science, if we must follow Western-movie scenarios of retrospective history). As the centerpiece of his logic, Burnet insists again and again that the earth's scripturally specified history will be adequately explained only when we identify natural causes for the entire panoply of biblical events. Moreover, he urges, in apparent conflicts (they cannot be real) between reason and revelation, choose reason first and then untangle the true meaning of revelation:

'Tis a dangerous thing to engage the authority of scripture in disputes about the natural world, in opposition to reason; lest time, which brings all things to light, should discover that to be evidently false which we had made scripture to assert . . . We are not to suppose that any truth concerning the natural world can be an enemy to religion; for truth cannot be an enemy to truth, God is not divided against himself. (16)

3. This commitment led Burnet into arguments that we, with different assumptions, might regard as the height of folly—that, for example, Noah's flood must have been truly global, not merely local, because Noah would not have built an ark, but simply fled to safety in a neighboring land, if the entire earth had not been drowned.

Burnet strenuously attacked those who would take the easy road
and call upon miraculous intervention whenever a difficult problem
presented itself to physics—for such a strategy cancels reason as a
guide and explains nothing by its effortless way of resolving everything. In rejecting a miraculous creation of extra water to solve the
central problem that motivated his entire treatise—how could the
earth drown in its limited supply of water?—Burnet invoked the
same metaphor later used by Lyell against the catastrophists: easy
and hard ways to untie the Gordian knot. "They say in short, that
God Almighty created waters on purpose to make the deluge, and
then annihilated them again when the deluge was to cease; And
this, in a few words, is the whole account of the business. This is
to cut the knot when we cannot loose it" (33). Likewise, for the
second greatest strain on physical credulity, a worldwide conflagration, Burnet again insists that ordinary properties of fire must do
the job: "Fire is the instrument, or the executive power, and hath
no more force given it, than what it hath naturally" (271).

Burnet's basic position has been advanced by nearly every theistic
scientist since the Newtonian revolution: God made it right the
first time. He ordained the laws of nature to yield an appropriate
history; he needn't intervene later to patch and fix an imperfect
cosmos by miraculous alteration of his own laws. In a striking
passage, Burnet invokes the standard metaphor of clockwork to
illustrate this primary principle of science—the invariance of natural
laws in space and time.

> We think him a better artist that makes a clock that strikes
> regularly at every hour from the springs and wheels which he
> puts in the work, than he that hath so made his clock that he
> must put his finger to it every hour to make it strike: and if one
> should contrive a piece of clock-work so that it should beat all
> the hours, and make all its motions regularly for such a time,
> and that time being come, upon a signal given, or a spring
> touched, it should of its own accord fall all to pieces; would not

this be looked upon as a piece of greater art, than if the workman came at that time prefixed, and with a great hammer beat it into pieces?   (89)

Only late in the book, when he must specify the earth's future following the conflagration, does Burnet admit that reason must fail—for how can one reconstruct the details of an unobservable future? Yet he abandons reason with much tenderness and evident regret:

Farewell then, dear friend, I must take another guide: and leave you here, as Moses upon Mount Pisgah, only to look into that land, which you cannot enter. I acknowledge the good service you have done, and what a faithful companion you have been, in a long journey: from the beginning of the world to this hour . . . We have travelled together through the dark regions of a first and second chaos: seen the world twice shipwrecked. Neither water nor fire could separate us. But now you must give place to other guides. Welcome, holy scriptures, the oracles of God, a light shining in darkness.   (327)

The Physics of History

I have already presented the content of Burnet's scenario in outline by discussing his frontispiece; but what physics did he invoke to produce such an astonishing sequence of events?

Burnet viewed the flood as central to his methodological program. The *Sacred Theory,* therefore, does not proceed chronologically, but moves from deluge to preceding paradise, for Burnet held that if he could find a rational explanation for this most cataclysmic and difficult event, his method would surely encompass all history. He tried to calculate the amount of oceanic water (Figure 2.2), grossly underestimating both the average depth (100 fathoms) and extent (half the earth's surface) of the seas.[4] Concluding that the seas could not nearly bury the continents, calculating that forty days and nights of rain would add little (and only recycle seawater in any case), and rejecting, as methodologically destructive to his rational program, the divine creation of new water, Burnet had to seek another source. He fixed upon a worldwide layer of water, underlying and concentric with the original crust of the earth's

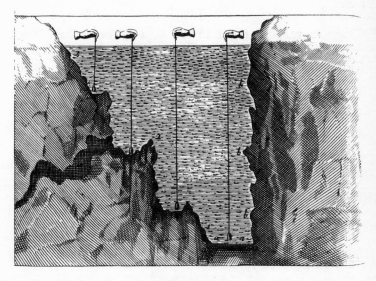

Figure 2.2
Burnet attempts to assess the amount of water in the oceans by the classical method of sounding. (From first edition.)

4. Burnet, who was not the armchair speculator of legend, lamented the absence of adequate maps to make assessments and calculations for these key elements of his theory: "To this purpose I do not doubt but that it would be of very good use to have natural maps of the earth . . . Methinks every prince should have such a draught of his own country and dominions, to see how the ground lies . . . which highest, which lowest . . . how the rivers flow, and why; how the mountains stand . . ." (p. 112).

surface. The flood, he declared, occurred when this original crust cracked, permitting the thick, underlying layer of water to rise from the abyss (Figure 2.3).

This interpretation of the flood allowed Burnet to specify conditions both before and after. Nothing much has happened since the deluge, only some inconsequential erosion of postdiluvian topography. (Burnet's geology lacked a concept of repair; processes of ordinary times could only follow the dictates of Isaiah 40 and erode the mountain to fill the valleys, thus smoothing and leveling the surface.)

The earth's current surface was fashioned by the deluge (Figure 2.4). It is, in short, a gigantic ruin made of cracked fragments from the original crust. Ocean basins are holes, mountains the edges of crustal fragments broken and turned upon their side. "Say but that they are a ruin, and you have in one word explained them all" (101). Burnet's descriptions and metaphors all record his view of our current earth as a remnant of destruction—a "hideous ruin," "a broken and confused heap of bodies," "a dirty little planet."

Burnet then proceeded backward (in Book II) to reconstruct the perfect earth before the deluge. Scripture specifies an original chaos of particles (Figure 2.5), and physics dictates their sorting as a series of concentric layers, denser at the center (Figure 2.6). (Since Burnet regarded the solid crust as a thin and light froth, denser water formed a layer beneath—and a source for the deluge.)

This perfect earth housed the original paradise of Eden. Its surface was featureless and smooth. Rivers ran from high latitudes and dissipated in the dry tropics (Figure 2.7). (They flowed, in Burnet's reversed concept of the earth's shape, because the poles stood slightly higher above the center than the equator.) A planet with such perfect radial symmetry bore no irregularity to tilt its axis. Hence the earth rotated bolt upright and Eden, located at a mid-latitude, enjoyed perpetual spring. The salubrious conditions of this earthly paradise nurtured the early patriarchal life-spans of more than nine hundred years. But the deluge was truly paradise lost. The earth, made asymmetric, tilted to its present angle of some

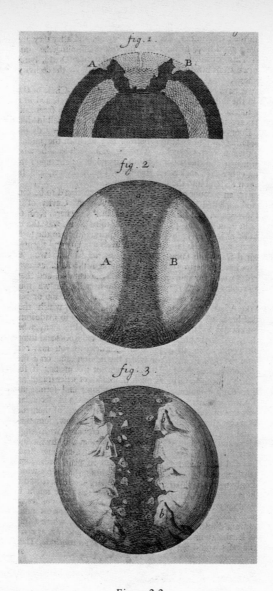

Figure 2.3
Burnet's physical cause of the deluge. The earth's crust cracks (fig. 1);
water rises from its layer in the abyss to cover the earth (fig. 2), and then
retreats again to leave modern continents (with mountains as edges of the
cracked crust) and oceans. (From first English edition.)

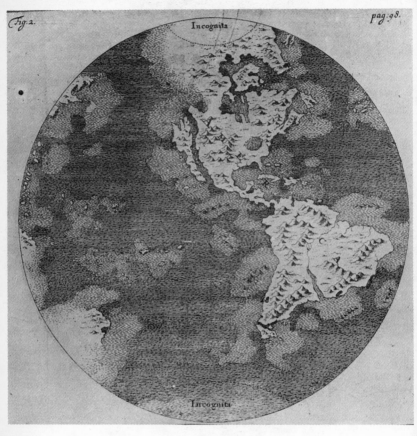

Figure 2.4
The earth's current surface, a product of crustal collapse
during the deluge. (From first edition.)

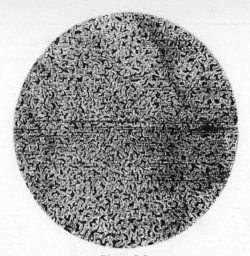

Figure 2.5
The chaos of the primeval earth as related in Genesis 1.
(From first edition.)

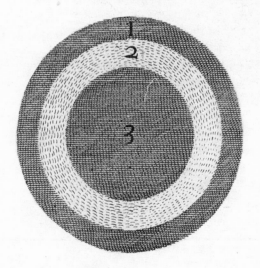

Figure 2.6
The perfect earth of the original paradise of Eden, arranged as concentric
layers according to density after the descent of particles from primeval
chaos. (From first edition.)

twenty degrees. An unhealthy change of seasons commenced, and life-spans declined to the currently prescribed three score years and ten.

If this reconstruction strikes modern readers as fanciful and scripture-burdened (I do not deny it, but only urge different criteria of judgment), recall Burnet's commitment to a rational explanation based on natural laws. We might contrast Burnet's account of the change in axial tilt with the words of a celebrated near-contemporary who did not shrink from attributing the work directly to angels:

> Some say he bid his angels turn askance
> The poles of Earth twice ten degrees and more
> From the sun's axle; they with labor pushed
> Oblique the centric globe: some say the sun
> Was bid turn reins from the equinoctial road
> Like distant breadth to Taurus with the seven
> Atlantic Sisters, and the Spartan Twins,
> Up to the Tropic Crab; thence down amain
> By Leo and the Virgin and the Scales,
> As deep as Capricorn, to bring in change
> Of seasons to each clime; else had the spring
> Perpetual smiled on Earth with vernant flowers,
> Equal in days and nights, . . .
> The sun, as from Thyestean banquet, turned
> His course intended; else how had the World
> Inhabited, though sinless, more than now
> Avoided pinching cold and scorching heat?

> Milton, *Paradise Lost*

In Book III Burnet presents a set of arguments, guided more by scripture since physics treats pasts more securely than futures, for a coming worldwide conflagration that shall completely consume the upper layers of the earth and remobilize all resulting particles into a new chaos. Burnet continues to demand a rational, physical explanation. In successive chapters, he discusses how such a wet and rocky mass can burn (the waters will first evaporate in a major drought), how the jumbled surface of our ruined earth will abet

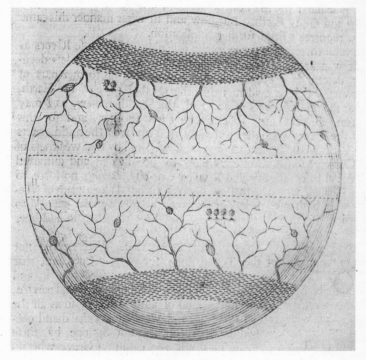

Figure 2.7

The earth's surface in its paradisiacal state. Rivers run from high latitudes and dissipate in the tropics. Four trees mark the position of Eden at a convenient middle latitude in the southern hemisphere. (From first English edition.)

the flames (by including so much nurturing air from internal va-cuities), and where the fire will start. The torches of Vesuvius and Aetna specify an Italian origin, and God also knows the home of antichrist, the Bishop of Rome (Burnet was nothing if not a com-mitted Anglican). Still, in an ecumenical spirit, Burnet tells us that Britain, with its deposits of coal, will burn brightly, if a bit later.

If the conflagration replays the flood in a different manner, the succeeding earth also reproduces the original paradise, and for the same physical reason: the descent of burned particles into concentric layers sorted by density (Figure 2.8). On this earth made perfect again, Christ shall reign for a thousand years with Satan bound in chains. Following this millennium, Gog and Magog shall herald the final battle of good against evil; trumpets shall announce the last judgment; the saints shall ascend (the sinners go elsewhere); and the superseded earth shall become a star.

Burnet's commitment to explanation by natural law, and his allegiance to historical narrative, can be best appreciated in explicit contrast with alternatives proposed by his friend Isaac Newton in a fascinating exchange of long letters (thank goodness they didn't have a telephone, or even a train) between London and Cambridge, in January 1681.[5]

Newton had made two suggestions that troubled Burnet: first, that the earth's current topography arose during its initial formation from primeval chaos, and was not sculpted by Noah's flood; second, that the paradox of creation in but six days might be resolved by arguing that the earth rotated much more slowly then, producing a "day" of enormous extent. Burnet excused his long and impas-sioned response, stating (in modern orthography):

Your kindness hath brought upon you the trouble of this long letter, which I could not avoid seeing you have insisted upon two such material points, the possibility (as you suppose) of

5. I thank Rhoda Rappoport for sending me these documents.

forming the earth as it now is, immediately from the Chaos or without a dissolution [as the flood had produced in Burnet's scheme]; and the necessity of adhering to Moses' Hexameron as a physical description [Burnet actually wrote "Moses his Hexameron" in that delightful construction used before English codified the current form of the genitive]; to show the contrary to these two hath swelled my letter too much.   (in Turnbull, 1960, 327)

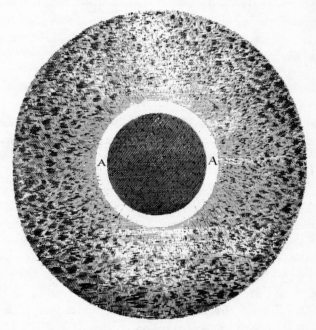

Figure 2.8
The earth made perfect a second time, after the descent of particles by density into concentric layers following the future conflagration.
(From first edition.)

Burnet objected to Newton's first proposal because it removed the role of extended history by forming all the earth's essential features at the outset—see quote above with its plea for the flood as an agent of topography. Burnet then rejected Newton's long initial days because he suspected that a later acceleration in rotation would require supernatural intervention. (Burnet favored an allegorical interpretation of Genesis 1, arguing that the notion of a "day" could not be defined before the sun's creation on the fourth day.) "But if the revolutions of the earth were thus slow at first, how came they to be swifter? From natural causes or supernatural?" (325). Burnet also objected that long early days would stretch the lives of patriarchs even beyond the already problematical 969 of Methuselah and his compatriots—and that while organisms might enjoy sunny days of such extended length, the long nights might become unbearable: "If the day was thus long what a doleful night would there be" (325).

Newton's response confirms Burnet's reading of their differences. Newton argues that a separation of parts from original chaos might produce irregular topography, not the smooth concentric layers of Burnet's system—therefore requiring no subsequent narrative to explain the current face of our earth: "Moses teaches a subdivision . . . of the miry waters under the firmament into clear water and dry land on the surface of the whole globous mass, for which separation nothing more was requisite than that the water should be drained from the higher parts of the limus to leave them dry and gather together into the lower to compose seas. And some parts might be made higher than others" (333).

As for an early speeding of rotation, Newton confirmed Burnet's fears by allowing a direct supernatural boost: "Where natural causes are at hand God uses them as instruments in his works, but I do not think them alone sufficient for the creation and therefore may be allowed to suppose that amongst other things God gave the earth its motion by such degrees and at such times as was most suitable to the creatures" (334). Newton also disregarded the problem of "doleful nights" for the earth's first inhabitants, arguing:

"And why might not birds and fishes endure one long night as well as those and other animals endure many in Greenland" (334).

Burnet therefore emerges from this correspondence with the greatest of all scientific heroes as more committed to the reign of natural law, and more willing to embrace historical explanations. He ends his letter to Newton by describing a singular event of time's arrow, the great comet of 1680 then hanging over the skies of London. "Sir we are all so busy in gazing upon the comet, and what do you say at Cambridge can be the cause of such a prodigious coma as it had" (327). Mr. Halley, mutual friend of Newton and Burnet, also gazed in awe at this comet. Two years later, still inspired by this spectacular sight, he observed a smaller comet, and eventually predicted its return on a seventy-six-year cycle. This smaller object, Halley's Comet, now resides in my sky as I write this chapter—a primary signal of time's cycle.

Time's Arrow, Time's Cycle: Conflict and Resolution

In focusing upon Burnet's rationalist methodology, revisionists who wish to identify something of value in his work have missed an important opportunity—in part, because they rely exclusively upon text and ignore pictures. I view Burnet's frontispiece as the finest expression ever published of the tension between two complementary views of time—the ancient contrast of time's arrow and time's cycle. I studied Burnet's text again, with this perspective drawn from his frontispiece, and understood it in a new light (after half a dozen previous readings). I saw the *Sacred Theory* as a playground for Burnet's struggle to combine the metaphors into a unified view of history that would capture the salient features of each—the narrative power of the arrow, and the immanent regularity of the cycle. I think that Burnet's struggle was quite conscious; his frontispiece is constructed with consummate care.

Burnet's Portrayal and Defense of Time's Arrow

A defense of narrative

Since the *Sacred Theory* is primarily a story, the concerns and metaphors of time's arrow dominate the text. Burnet locates his rationale for writing in a desire to establish the idea of directional history against the Aristotelian notion of changeless or cycling eternity. Before launching into his narrative, Burnet devotes a chapter to refuting the Aristotelian premise. He presents some theoretical arguments for a beginning of earthly time, then pauses and realizes that his coming narrative will serve as ample refutation of nondirectional eternity: "We need add no more here in particular, against this Aristotelian doctrine, that makes the present form of the earth to have been from eternity, for the truth is, this whole book is one continued argument against that opinion."

Again and again, Burnet justifies his attention to sequential narrative as an approach to understanding the earth. It is, first of all, fun: "I had always, methought, a particular curiosity to look back into the first sources and original of things; and to view in my mind . . . the beginning and progress of a rising world" (23). "There is a particular pleasure to see things in their origin, and by what degrees and successive changes they rise into that order and state we see them in afterwards, when completed" (54).

We require narrative for several reasons, including the natural proclivities of human curiosity: "Tis natural to the mind of man to consider that which is compound, as having been once more simple" (43); the dictates of reason: "There is no greater trial or instance of natural wisdom, than to find out the channel, in which these great revolutions of nature, which we treat on, flow and succeed one another" (66); the ways of divinity: "I am sure, if ever we would view the paths of Divine Wisdom, in the works and in the conduct of nature, we must not only consider how things are, but how they came to be so" (54); and the sheer magnitude of history's impact: "I am apt to think that some two planets, that are under the same state or period, do not so much differ from one another, as the same planet doth from itself, in different periods of its duration. We do not seem to inhabit the same world that our first fore-fathers did, nor scarce to be the same race of men" (140).

What features of the earth compel us to explain it by narrative?

In seeking criteria to argue that a current state demands explanation as a product of history, Burnet occasionally invokes the evidence of direct observation: "If one should assert that such an one has lived from all eternity, and I could bring witnesses that knew him a sucking child, and others that remembered him a school-boy, I think it would be fair proof, that the man was not eternal."

But only rarely can we observe enough change directly, for ancient (particularly antediluvian) texts do not exist, and witnesses are mute. We must therefore seek a criterion of inference from present states or the surviving artifacts of a different past.

Burnet champions the same resolution that Darwin would invoke 150 years later as he struggled to interpret the shapes of organisms as products of history.[6] Darwin gave his answer in paradoxical form: history lies revealed in the quirks and imperfections of modern structures. Evolved perfection covers the tracks of its own formation. Optimal designs may develop historically, but they may also be created *ab nihilo* by a wise designer. What Darwin advanced for the bodies of organisms, Burnet applied to the form of the earth. The current earth is a ruin, "shapeless and ill-figured" (112). But a ruin can only be a wreck of something once whole, in short a product of history. The earth's current form makes sense only as one stage of a developing pageant:

> There appearing nothing of order or any regular design in its parts, it seems reasonable to believe that it was not the work of nature, according to her first intention, or according to the first model that was drawn in measure and proportion, by the line

6. I do not make this comparison to praise Burnet because he anticipated a later hero, but only to maintain that fine thinkers face similar problems across the centuries and often apply the same rules of fruitful reason.

and by the plummet, but a secondary work, and the best that could be made of broken materials.   (102)

The same principle regulates human artifacts, or anything built by wisdom, earthly or divine:

When this idea of the earth is present to my thoughts, I can no more believe that this was the form wherein it was first produced, than if I had seen the Temple of Jerusalem in its ruins, when defaced and sacked by the Babylonians, I could have persuaded myself that it had never been in any other posture, and that Solomon had given orders for building it so.   (121)

The treatment of narrative in form and metaphor

Since Burnet's *Sacred Theory* is, above all, a good story, narrative style predominates throughout. Burnet's presentation of the earth's essential historicity takes two forms. He argues, first, for *vectors*—patterns of directed order and definite duration. The earth, for example, cannot be eternal because erosion is a one-way street that eventually destroys all topography:

If this present state and form of the earth had been from eternity, it would have long ere this destroyed itself . . . the mountains sinking by degrees into the valleys, and into the sea, and the waters rising above the earth . . . For whatsoever moulders or is washed away from them is carried down into the lower grounds, and into the sea, and nothing is ever brought back again by any circulation. Their losses are not repaired, nor any proportionable recruits made from any other parts of nature.   (44–45)

Second, and most often, Burnet simply indulges his skill for pure *narrative*—a story line of pasts that determine presents and presents that constrain futures. We could dip almost anywhere into his text, but consider just this description of the deluge:

Upon this chaos rid the distressed ark, that bore the small remains of mankind. No sea was ever so tumultuous as this, nor is there anything in present nature to be compared with the disorder of these waters; all the poetry and all the hyperboles that are used in the description of storms and raging seas, were literally true in this, if not beneath it. The ark was really carried to the tops of the highest mountains, and into the places of the clouds, and thrown down again into the deepest gulfs. (84)

Burnet's closing words again identify his treatise as a work of historical narrative: "There we leave [the earth]; having conducted it for the space of seven thousand years, through various changes from a dark chaos to a bright star" (377).[7]

Since Burnet was a fine prose stylist, we can also trace his primary commitments by recording his metaphors. The direction, or vector, of history is a "channel" (66) or "progress" along "the line of time" (257); while the narrative quality of history receives, over and over again, its evident analogy with theater: "To see, when this theater is dissolved, where we shall act next, and what parts. What saints and heroes, if I may so say, will appear upon that stage; and with what luster and excellency" (241).

Burnet unites his criteria of vector and narrative in another obvious analogy: the history of our planet with the growth of a tree. This passage also includes his finest defense of the beauty and necessity of historical analysis:

We must not only consider how things are, but how they came to be so. 'Tis pleasant to look upon a tree in the summer, covered with its green leaves, decked with blossoms, or laden with fruit, and casting a pleasing shade under its spreading boughs; but to consider how this tree with all its furniture, sprang from a little seed; how nature shaped it, and fed it, in its infancy and growth; added new parts, and still advanced it by little and little, till it came to this greatness and perfection, this, methinks, is another sort of pleasure, more rational, less common . . . So to view this

7. Note how Burnet incorporates both features of history—the vector of direction (dark chaos to bright star) and the rich pageant of narrative (various changes).

earth, as it is now complete, distinguished into the several orders of bodies of which it consists, every one perfect and admirable in its kind; this is truly delightful, and a very good entertainment of the mind; but to see all these in their first seeds, as I may so say; to take in pieces this frame of nature, and melt it down into its first principles; and then to observe how the divine wisdom wrought all these things out of confusion into order, and out of simplicity into that beautiful composition we now see them in; this, methinks, is another kind of joy, which pierceth the mind more deep, and is more satisfactory.   (54)

Burnet's Portrayal and Defense of Time's Cycle

The *Sacred Theory* is much more than pure narrative. It tells a story within a prescribed framework. Time cannot simply move forward toward ever more different and progressive states. God, and nature's order, forbids a mere aimless wandering through time's multifarious corridors. Our modern earth separates two grand cycles of repetition—our past and future. Destruction (deluge) followed perfection (paradise) in our past. Our future shall cycle through these same stages in reverse order, and with uncanny precision of detail—destruction (conflagration) *before* renewed perfection.

If reason implies an order of exquisite design (as Burnet had argued in describing the earth's original perfection), then the fabric of time as a whole must display rational order as well—for God superintends both space and time. Such rational order demands an immanent timelessness of invariant, or cyclically repeating, pattern. Thus, time's cycle pervades the *Sacred Theory* as surely as time's arrow. The arrow moves forward within a framework of repetition that forms the signature of inherent order and good sense in the cosmos.

At the very end of Book IV, Burnet argues that cycles are the way of God and nature: "Revolution to the same state again, in a great circle of time, seems to be according to the methods of Providence; which loves to recover what was lost or decayed, after certain periods: and what was originally good and happy, to make it so again" (376). (Burnet uses the word "revolution" in its Newtonian, pre-1776 and pre-Bastille sense to mean turning, not upheaval.)

Cycles also embody an aesthetic necessity, for the world would be impoverished and ill-formed without a concept of renewal for those parts of nature that wear out:

> There would be nothing great or considerable in this inferior world, if there were not such revolutions of nature. The seasons of the year, and the fresh productions of the spring, are pretty in their way; but when the Great Year comes about, with a new order of all things, in the heavens and on the earth; and a new dress of nature throughout all her regions, far more goodly and beautiful than the fairest spring; this gives a new life to the creation, and shows the greatness of its author.   (246)

Burnet pays special homage to time's cycle in discussing the recovery of paradise following the conflagration. The renewal of perfection shall be so precise that only an inherent cyclicity of time could underlie such a restoration—for topography shall be smooth again, and by the same method (sorting by density of a chaotic mass of particles into concentric layers); while restored radial symmetry shall rectify the earth's axis to its original upright position. Thus, all parts of the globe shall "be restored to the same posture they had at the beginning of the world; so as the whole character of the Great Year would be truly fulfilled . . . A general harmony and conformity of all the motion of the universe would presently appear, such, as they say, was in the Golden Age, before any disorder came into the natural or moral world" (257).

As Burnet developed a suite of metaphors to capture time's arrow, his discussions of time's cycle invoked a set of analogies to capture the other face of history. Some are geometric (contrasting with the "line" or "channel" of time's arrow)—"the circle of successions"; "the great circle of time and fate" (13). Others invoke the seasonal and annual cycles of our own experience (contrasting with analogies

to the human life-span for time's arrow)—"a new dress of nature
... more beautiful than the fairest spring" (246); or, especially, the
annus magnus or "great year" of geological time. The twin themes
of time's cycle are revolution and restoration.

The Resolution of Advancing Cycles

These two discussions seem to mark an irreconcilable conflict. How
could Burnet exalt both the necessity of an arrow to identify history
(see page 32 on the vector of erosion) and a cycle to record divine
superintendence (see page 46 on the necessity of restitution, the
very point apparently denied in describing the vector of erosion).

Burnet's own resolution[8] of this dilemma, and his union of these
two superficially contrary themes, begins by describing the dilemma
that a purely cyclical view, without arrows, entails. Attacking the
ancient Greek notion of exact cyclic repetition, he states: "They
made these revolutions and renovations of nature, indefinite or
endless: as if there would be such a succession of deluges and
conflagrations to all eternity" (249). Burnet recognizes that such a
vision destroys the very possibility of history—"this takes away the
subject of our discourse," he states (43)—for no event can be placed
into narrative if each occurred before and must happen again. This
dilemma—the incomprehensibility of infinity—was beautifully expressed by Jorge Luis Borges in "The Book of Sand." In this story,
Borges trades his precious Wycliffe Bible for an amazing, infinite
book. One cannot find its beginning, for no matter how furiously
you flip the pages, there are always just as many between you and
the front cover; the book has no end for the same reason. It contains
small illustrations, two thousand pages apart, but none is ever
repeated, and Borges soon fills his notebook with a list of their
forms, coming no closer to any termination. Finally, Borges realizes

8. Burnet did not invent this argument. He presents here the traditional resolution, favored both before his time (as I shall discuss in the last chapter on medieval
iconography) and after (as in the Hegelian, and later Marxist, conception of cycles
moving onward—negation of negation to new, not merely repetitive, states).

that the book is monstrous and obscene—and he loses it permanently on a shelf in the stacks of the Argentine National Library.

The trader's epitome of this impossible book presents the same dilemma that Burnet avoids by rejecting the strict cycle and arguing for narrative: "If space is infinite, we may be at any point in space. If time is infinite, we may be at any point in time."

If the arrow by itself makes time unintelligible, but if nature is formless, and therefore incomprehensible, without cycles, then what is to be done? Burnet argues that cycles must turn, but that phases repeat with crucial differences each time. The material substrate does not change (for the same stuff cycles), but resulting forms alter, often in a definite direction so that each repetition passes with distinctive and identifiable differences. We can, in other words, know where we are—and Borges's paradox is resolved. Burnet returns to Aristotle's error and illustrates, with his favorite metaphor of the stage, why we must incorporate elements of directional change at each repetition. The theorists of time's cycle assert

> the identity, or sameness, if I may so say, of the worlds, succeeding one another. They are made indeed of the same lump of matter, but they supposed them to return also in the same form . . . So as the second world would be but a bare repetition of the former, without any variety or diversity . . . As a play acted over again, upon the same stage, and to the same auditory. (249)

Burnet then brands time's cycle in its pure form as "a manifest error . . . easily rectified" (249). Both nature and scripture impose a vector of history upon any set of cycles, guaranteeing that no phase can repeat exactly the corresponding part of a former cycle:

> For, whether we consider the nature of things; the earth after a dissolution; by fire or by water, could not return into the same form and fashion it had before; or whether we consider providence, it would in no ways suit with the divine wisdom and justice to bring upon the stage again those very scenes, and that

very course of human affairs, which it had so lately condemned
and destroyed. We may be assured therefore, that, upon the
dissolution of a world, a new order of things, both as to nature
and providence, always appears.   (249)

Burnet's discussion of the second, or future, cycle weaves a subtle
interplay between elements of repetition (to display order and plan),
and strands of difference (to permit a recognizable history). He
stresses the detailed similarity between two destructions that seem
so different—the deluges (as he calls them both) of water and fire.
Both were global and both required a union of agents from above
and below the earth's surface: rain from above meeting an upwelling
layer of water breaking through the crust, and lightning above
joining with subterranean flames and lavas of exploding volcanoes:
"There is a great analogy to be observed betwixt the two deluges,
of water and of fire; not only as to the bounds of them . . . but as
to the general causes and sources upon which they depend, from
above and from below" (277).

But the changes of history create uniqueness within these striking
similarities. The details of narrative make each repetition a separate
story. Fire and water destroy differently: "The earth must be re-
duced into a fluid mass, in the nature of a chaos, as it was at first;
but this last will be a fiery chaos, as that was watery; and from this
state it will emerge again into a paradisiacal world" (288).

Burnet wishes to discern more than simple difference from cycle
to cycle. He tries to detect a *vector* of progress as well. Destruction
by fire must yield something even better than the first paradise; the
cycles roll onward. Burnet claims that fire can purge and purify
more fully than any process operating during the first cycle of the
earth's past: "Nature here repeats the same work, and in the same
method; only the materials are now a little more refined and purged
by fire" (324).

We now see why Burnet's frontispiece captures the essence of his
system so clearly and succinctly. It shows the detailed correspondences between present and future, the two great cycles of our

planet's course; but history moves inexorably forward as it cycles. Dark chaos lies under Christ's left foot, marking our beginnings; but the bright star of our ending closes the circle of the great year.

Burnet and Steno as Intellectual Partners in the Light of Time's Arrow and Time's Cycle

Since we use individuals to illustrate general attitudes, Burnet has been forced to play another, unhappy role in textbook histories. He becomes the symbol of constraining bibliolatry, with its Mosaic time-scale and miraculous paroxysms. He stands in traditional contrast with his contemporary Nicolaus Steno, the great Danish savant (later Bishop, in a conversion generally viewed as anomalous and retrogressive) whose *Prodromus* of 1669 marks the conventional inception of modern geology. Steno succeeded, we are told, because he adopted the universal procedures of scientific method so respected today. The introduction to the standard translation of the *Prodromus* calls Steno "a pioneer of the observational methods which dominate in modern science" (Hobbs, 1916, 169). The translator adds: "At a time when fantastic metaphysics were rife, Steno trusted only to induction based upon experiment and observation" (Winter, 1916, 179).

Burnet then becomes the symbol of those "absurdities of metaphysical speculation" (Winter, 1916, 182)—in explicit contrast with Steno, who saved science before switching to souls. Thus, we learn from Fenton and Fenton (1952, 22) that Burnet added nothing to science, "nor compared with Steno" as a general thinker. While Davies descends from sublimity to bathos with these words: "From the wisdom of Steno, we must now turn to the fanciful, but ingenious and extremely popular theory of the earth devised by Thomas Burnet" (1969, 68).

Yet as I reread Steno in the light of my thoughts on metaphors of time, I realized that a set of remarkable similarities, hidden by later traditions of discourse, unite the scriptural panorama of Burnet with the observational discourse of Steno. I now believe that the similarities are more significant than the obvious differences.[9]

To grasp these similarities, we must focus on the part of the *Prodromus* that textbook commentators find embarrassing and usually neglect or mention with apology (Hobbs, 1916, 170, calls it a "weak conclusion intended to prove the orthodoxy of his position")—the last section, part four, on the geological history of Tuscany. Critics find this part disappointing because Steno both asserts the conformity of his geological history with the events of scripture and states his allegiance to the short Mosaic chronology (263). (These positions, of course, represent an important similarity with Burnet, but not the concordance that I wish to emphasize.)

We may best grasp Steno's scheme by turning again to a picture—his epitome, in six stages, of the history of Tuscany. I present it in the only form available to most commentators (Figure 2.9)—the reproduction in the standard translation (Winter, 1916) and at least two major histories (Chorley et al., 1964, 10; Greene, 1961, 60).

In this version, a vertical sequence of six stages, we can readily view Steno's history as usually presented—an ordinary time's-arrow approach to the earth, telling a story of sequential events leading in a definite direction. We see the original state of the earth in figure 25, covered with water and receiving sediments in a universal basin. (Strictly speaking, Steno records only the history of Tuscany here—but he states that the whole earth follows the same pattern: "As I prove this fact concerning Tuscany by inference from many places examined by me, so do I affirm it with reference to the entire earth, from the descriptions of different places contributed by different writers," 263.) Underground waters and internal movements of the earth then create vacuities within the pile of sediments,

9. I do not wish to belittle the central distinction that Burnet reports few personal (and no field) observations, while Steno describes mineral specimens and discusses the lay of the Tuscan landscape—though in terms so broad and general that I find no evidence for extensive fieldwork in any modern sense. This difference became crucial in the light of later developments, but the neglected similarity of theoretical structure has much insight to offer as well.

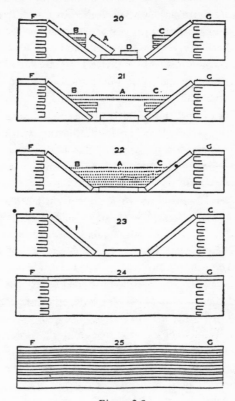

Figure 2.9
Steno's geological history of Tuscany, as rearranged in the English translation by J. G. Winter (1916), presumably to conform with the later geological tradition (but not Steno's concept) of time as a linear sequence.

leaving the upper layer intact but excavating the strata below to yield a precarious instability (figure 24). The crust collapses into the vacuities (figure 23) and the waters of Noah's flood rise to deposit another pile of sediments within the newly formed basin (figure 22). During the subsequent calm after the flood, vacuities arise again within the new pile of sediments (figure 21), and the crust of flood deposits finally collapses into the interior spaces, leaving the earth as we find it today (figure 20).

Steno's allegiance to time's arrow pervades more than his figures. He explores and defends this general approach to history at length—and with the same arguments that Burnet used. Steno agrees with Burnet that a reconstruction of the earth's past must rest upon signs of history preserved in the current form of objects. We must learn (262) "in what way the present condition of any thing discloses the past condition of the same thing" (*quomodo praesans alicuius rei status statum praeteritum eiusdem rei detegit*). As his chief criterion, Burnet proposed the disorder of the earth's surface as a sure sign of change from original perfection. Steno uses the same argument. His original earth is a smooth pile of sediments, and their disruption provides his explicit criterion of history: "inequalities of surface observed in its appearance today contain within themselves plain tokens of different changes" (262).

Had I relied upon the available published figure, I might never have grasped the richness of Steno's conception or discovered its similarities with Burnet's. But one day, for aesthetic pleasure and as a sort of sacrament, I turned through all pages of an original copy of the *Prodromus*. When I came to Steno's own version of Tuscan history (Figure 2.10), I immediately noted the difference from later reproductions, but I did not at first appreciate the significance of this alteration.

Steno did not group the six stages in one vertical column, but in two sets of three, one beside the other. Later historians have rearranged the figure according to modern conventions, with geological time flowing in one direction as a set of strata, oldest at the bottom. I can only suppose that they accorded no special significance to Steno's choice—for none have commented upon the reorientation. Perhaps they assumed that Steno placed his figures so oddly (by modern convention) because he had to fit all six on the lower part of one sheet, below several figures of crystals—and had no room for a single vertical column. Perhaps, since pictures usually receive such short shrift as historical documents, they hardly considered the issue at all, and simply arranged the figure as any modern geologist would.

I stored this difference in the back of my mind, and appreciated its significance only when I started thinking seriously about relationships between metaphors of time. The rearranged vertical column is time's arrow of modern convention. It reflects our dominant metaphor of time, and seems so "right" that we dismiss Steno's own orientation as inconsequential, or as an odd arrangement constrained by space available. But a study of Steno's text, and a general appreciation of how early geologists treated time's arrow *and* cycle, proves that Steno presented his version on purpose.

In telling the story of Tuscany, Steno did not confine his attention to narrative. He also sought to develop a cyclical theory of the earth's history. The six tableaux form two cycles of three, not a linear sequence. Each cycle passes through the same three stages: deposition as a uniform set of layers, excavation of vacuities within the strata, and collapse of the top strata into the eroded spaces,

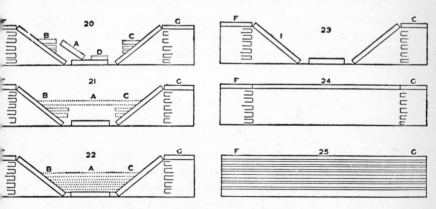

Figure 2.10
The geological history of Tuscany in Steno's original version, arranged
as two parallel columns.

producing a jumbled and irregular surface from an original smoothness. Steno's text makes this interpretation of *his* picture even more clear. His narrative may treat the six tableaux sequentially, but note his careful repetition of words for corresponding stages of the two cycles. For the second deposition of strata (shown in figure 22): "At the time when the plane BAC was being formed, and other planes under it, the entire plane BAC was covered with water; or what is the same thing, the sea was at one time raised above sand hills, however high" (262–263). And for the first deposition of strata, figure 25 (Steno recounts his narrative from youngest to oldest, in reverse order from current convention): "When the plane FG was being formed, a watery fluid lay upon it; or, what is the same thing, the plane summits of the highest mountains were at one time covered with water" (263).

In his summary statement, Steno writes primarily of repetition, not sequence: "Six distinct aspects of Tuscany we therefore recognize, two when it was fluid, two when level and dry, two when it was broken" (263).

Moreover, Steno does not restrict his defense of cyclic history to the data of Tuscany. He also presents a general argument for cyclic repetition as an inherent property of time and natural processes: "It cannot be denied that as all the solids of the earth were once, in the beginning of things, covered by a watery fluid, so they could have been covered by a watery fluid a second time, since the changing of things of nature is indeed constant, but in nature there is no reduction of anything to nothing" (265).

But the similarities with Burnet go further. Burnet also presented a powerful argument for the proper integration of his two metaphors, based on what I have called Borges's dilemma of infinity. Steno resolves the potential tension in the same manner—by arguing for repetition *with a difference*. He also uses the same two criteria for assessing these differences: narrative and vector.

Steno even follows Burnet in insisting that differences between cycles must exist in order to make time intelligible—for otherwise, each new cycle repeats the last sequence precisely, and we can never know where we stand in the march of history. Thus, Steno (who secured his highest reputation for writing so brilliantly about the nature of sedimentation) states that if strata of the second cycle do not differ internally from beds of the first sequence, we shall possess no criterion for separating the cycles except superposition, or the discovery of one set directly atop the other. (And, as every geologist knows, our imperfect sedimentary record rarely provides such direct evidence. We must be able to identify each cycle from internal evidence, since we usually find rocks of only one cycle in any single place.)

Steno then presents an internal standard using Burnet's criterion of narrative. He notes that strata of the first cycle contain no fossils (as products of the original earth, before creation of life), while strata of the second cycle include remains of plants and animals destroyed in the flood. "Thus we must always come back to the fact that at the time when those strata [of the first cycle] were being formed, the rest of the strata did not yet exist . . . All things were covered by a fluid free from plants and animals, and other solids" (264).

Steno then invokes Burnet's second criterion of history—the search for vectors. He identifies two vectors as imparting a direction to time. First, each new sequence of strata becomes more restricted in geographic extent since it forms *within* the spaces left by collapsing crust of the last cycle. The first set covered the entire earth (lower right of Figure 2.10); the second (lower left) only fills the center.[10] Second, each successive collapse makes the earth's surface more uneven (upper right of the first cycle compared with upper left of the second), and we can know where we stand in time's arrow by progressive departure from original smoothness: "The very surface of the earth was less uneven, because nearer to its beginnings" (267).

We may, therefore, summarize the detailed similarities between Burnet and Steno, two geological theories usually viewed not only

10. A modern geologist might object that such a progressive restriction would soon reduce space for strata to nothing on an ancient earth. But Steno's earth was a young planet, but two cycles old, and of limited potential duration.

as maximally different but as almost ethically opposed as the very best and worst of their times:

1. Both use conjoined evidence of nature and scripture to narrate the earth's history. Steno's words on this dual methodology differ little from Burnet's, previously cited: "In regard to the first aspect of the earth, scripture and nature agree in this, that all things were covered by water; how and when this aspect began, and how long it lasted, nature says not, scripture relates" (Steno, 263).

2. The geological systems of both men invoke the same basic mechanics—a theory of crustal collapse (probably developed from Descartes). Both view the earth as originally smooth and concentric. Both interpret the flood as a union of waters rising from a concentric layer beneath the crust with a prolonged scriptural rain from above. Both argue that the flood shaped current topography by causing the collapse of an originally smooth crust into spaces left by rising waters. Both lack a concept of repair and tell the earth's history as a departure from original smoothness to greater irregularity.

3. Most important, Burnet and Steno read the earth's history as a fascinating mixture of time's arrow and time's cycle. Both present pictorial epitomes that forge these apparently conflicting metaphors to an interesting union. Burnet shows a circle of earths, rather than a line. Steno draws two parallel rows, rather than a single sequence (though later historians reordered his figure and blotted out this basic feature). For both, history turns as a set of cycles (time's cycle), but each repetition must be different (time's arrow), in order to make time intelligible by imparting a direction to history.

As a final comment, stated with almost cryptic brevity here since I shall return at length in the last chapter, each of time's metaphors embodies a great intellectual insight. Time's cycle seeks immanence, a set of principles so general that they exist outside time and record a universal character, a common bond, among all of nature's rich particulars. Time's arrow is the great principle of history, the statement that time moves inexorably forward, and that one truly cannot step twice into the same river. History grants absolute uniqueness *in toto*, although timeless principles may regulate parts and abstractions.

Time's arrow must also destroy in due time the numerically rigid schemes of both Burnet and Steno—a definite number of cycles with predictable repetition of stages. A planet's history could obey Burnet's circle or Steno's parallel lines only if a wise agent of order governed the cosmos and established laws that make the products of history unfold in such simple and rigorous patterns. Darwin's idea of a truly contingent history—time's arrow in its fullest sense— a quirky sequence of intricate, unique, unrepeatable events linked in a unidirectional chain of complex causes (and gobs of randomness), has made the numerological schemes of Burnet and Steno inconsistent with any acceptable notion of earthly time. We can offer no more profound refutation than "just history" to those who would seek such simple order in the products of time. But time's cycle has its own, continuing power as well—though we must seek it in other guises.

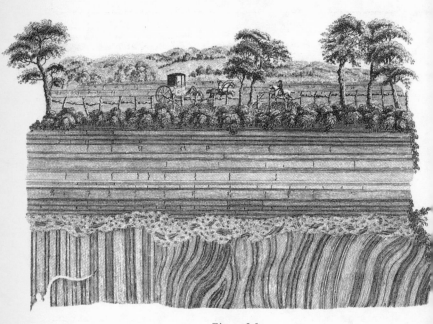

Figure 3.1
John Clerk of Eldin's celebrated engraving of Hutton's unconformity
at Jedburgh, Scotland.

CHAPTER THREE

James Hutton's Theory
of the Earth: A Machine
without a History

Picturing the Abyss of Time

The world is so complex, and the skills needed to apprehend it so varied, that even the greatest of intellects often needs a partner to supply an absent skill. As many of history's great lovers secured deputies to match physical appearance with the beauty of their poetry (the tragedy of Cyrano, among others), some scientists have needed a Boswell to present brilliant ideas in comprehensible form. James Hutton, whose *Theory of the Earth* (1795) marks the conventional discovery of deep time in British geological thought, might have occupied but a footnote to history if his unreadable treatise had not been epitomized by his friend, and brilliant prose stylist, John Playfair, in *Illustrations of the Huttonian Theory of the Earth* (1802).

In a familiar literary passage, Playfair described a great geological discovery that Hutton had showed him in 1788—not a thing so much as an interpretation. Hutton had recognized what we now call an unconformity as the most dramatic field evidence for time's vastness. Playfair described a phenomenon that Hutton would later depict in one of the few illustrations of his treatise, valued and reproduced ever since as a turning point in human knowledge. (It is, for example, the frontispiece both to this chapter—Figure 3.1— and to John McPhee's *Basin and Range*):

On us who saw these phenomena for the first time, the impression made will not easily be forgotten . . . We often said to ourselves, What clearer evidence could we have had of the different formation of these rocks, and of the long interval which separated their formation, had we actually seen them emerging from the bosom of the deep . . . Revolutions still more remote appeared in the distance of this extraordinary perspective. The mind seemed to grow giddy by looking so far into the abyss of time.

An unconformity is a fossil surface of erosion, a gap in time separating two episodes in the formation of rocks. Unconformities are direct evidence that the history of our earth includes several cycles of deposition and uplift.

I still use Hutton's drawing in my introductory courses to illustrate a principle that continues to stun me with its elegance after twenty annual repetitions—the complex panorama of history that can be inferred from the simple geometry of horizontal above vertical, once you understand the basic rules for deposition of strata. I can list a dozen distinct events that must have occurred to produce this geometry, with Hutton's unconformity as the key.

Since large expanses of water-laid strata must be deposited flat (or nearly so), the underlying vertical sequence arose at right angles to its current orientation. These strata were then broken, uplifted, and tilted to vertical in forming land above the ocean's surface. The land eroded, producing the uneven horizontal surface of the unconformity itself. Eventually, the seas rose again (or the land foundered) and waves further planed the old surface, producing a "pudding stone" of pebbles made from the vertical strata. Under the sea again, horizontal strata formed as products of the second cycle. Another period of uplift then raised these rocks above the sea once more, this time not breaking or tilting the strata. (Hutton reminds us that we must infer a second episode of uplift by drawing a meeting of phaeton and solitary horseman above the horizontal set of originally *marine* strata.) Thus, we see in this simple geometry of horizontal above vertical two great cycles of sedimentation with two episodes of uplift, the first tumultuous, the second more gentle.

Students have no trouble grasping this extended inference, and they do appreciate the point. Harder to convey is the revolutionary concept embedded in this inferred history, for Hutton's work helped to incorporate it among the commonplaces of modern thought. The revolution lies in a comparison with previous geological theories that included no mechanism for uplift and viewed the history of our planet as a short tale of uninterrupted erosion, as the mountains of an original topography foundered into the sea. This debate did not pit biblical idolatry against scientific thinking, as so often misportrayed; for as I noted in the last chapter, Steno's mechanical view shared with other seventeenth-century geologies the theme of continuous erosion as the organizing principle of history.

The pivot of debate was, instead and again, time's arrow and time's cycle. Hutton, I will argue, did not draw his fundamental inferences from more astute observations in the field, but by imposing upon the earth, *a priori,* the most pure and rigid concept of time's cycle ever presented in geology—so rigid, in fact, that it required Playfair's recasting to gain acceptability. Playfair aided Hutton's victory by soft-pedaling the uncompromising and ultimately antihistorical view of his late and dear friend.

In any case, this picture, and the unconformity that it represents, gains its cardinal significance as the primary item of *direct* evidence for time's cycle and an ancient earth. One can present abundant theory (as Hutton did) for the role of heat in uplifting strata, but unconformities are palpable proof that the earth does not decline but once into ruin; instead, by following decay with uplift, time cycles the products of erosion in a series that shows, in Hutton's most famous words, "no vestige of a beginning,—no prospect of an end" (1788, 304).

Hutton's World Machine and the
Provision of Deep Time

James Hutton had the good fortune to live in one of those rare conjunctions of time and place which, in a world not overpopulated with genius, bring a critical mass together for common purposes.

I would trade all the advantages of humanity to be a fly on the wall when Franklin and Jefferson discussed liberty, Lenin and Trotsky revolution, Newton and Halley the shape of the universe, or when Darwin entertained Huxley and Lyell at Down. Hutton was wealthy enough to be a full-time intellectual when Edinburgh was the thinking capital of Europe, and when David Hume, Adam Smith, and James Watt graced its supper clubs. Hutton belonged to that extinct species of eighteenth-century thinker, the polymath who took all knowledge for his province and moved with erudition between philosophy and science (a distinction that Hutton's contemporaries did not recognize). Most of Hutton's general treatises remain unpublished, and we have little idea of the extent of his vision. But he devoted most attention to developing a theory of the earth as a physical object existing for a purpose. He therefore becomes, under modern taxonomies, a geologist—and by modern mythologies, the father of geology.

Hutton has achieved this paternity because an English tradition[1] has claimed him as the primary discoverer of deep time—and all geologists know in their bones that nothing else from our profession has ever mattered so much.

For a historian, Hutton's rambling style provides the virtue of redundancy. You can always tell what he regarded as important because he says it over and over again. Few themes so pervade the thousand pages of his *Theory of the Earth* as his wonder and conviction about time's vastness. This subject even extracted some beautifully crafted and memorable lines from a man renowned (unfairly, I think) as the all-time worst writer among great thinkers. Consider the two most famous statements:

> Time, which measures every thing in our idea, and is often deficient to our schemes, is to nature endless and as nothing. (1788, 215)

1. I speak of language, not nationality. I know that Scotland isn't England, and don't wish to write British for a common English-language tradition of interpretation.

And the ringing, final line of the 1788 treatise:

> If the succession of worlds is established in the system of nature,
> it is in vain to look for anything higher in the origin of the earth.
> The result, therefore, of our present enquiry is, that we find no
> vestige of a beginning,—no prospect of an end.   (304)

The role of deep time within the mechanics of Hutton's theory
has been discussed ably and often, and I shall give only the briefest
outline here: By recognizing the igneous character of many rocks
previously viewed as sediments (products of decay), Hutton incor-
porated a concept of repair into geological history. If uplift can
restore an eroded topography, then geological processes set no limit
upon time. Decay by waves and rivers can be reversed, and land
restored to its original height by forces of elevation. Uplift may
follow erosion in an unlimited cycle of making and breaking.

Hutton describes the earth as a machine—a device of a particular
kind. Some machines wear out as their parts fall into irreversible
disrepair. But Hutton's world machine worked in a particular way
that prevented any aging. Something had to initiate the system (an
issue beyond the bounds of science, in Hutton's view), but once
kicked into action, the machine could never stop of its own accord,
because each stage in its cycle directly caused the next. As Playfair
wrote: "The Author of nature has not given laws to the universe,
which like the institutions of men, carry in themselves the elements
of their own destruction. He has not permitted, in his works, any
symptom of infancy or of old age, any sign by which we may
estimate either their future or their past duration" (1802, 119).

Hutton's self-renewing world machine works on an endlessly
repeating, three-stage cycle. First, the earth's topography decays as
rivers and waves disaggregate rocks, forming soils on the continents
and washing the products of erosion into the oceans. Second, the
comminuted bits of old continents are deposited as horizontal strata
in the ocean basins. As the strata build up, their own weight
generates sufficient heat and pressure to mobilize the lower layers.
Third, the heat of melting sediments and intruding magmas causes

matter to expand "with amazing force" (1788, 266), producing extensive uplift and generating new continents at the sites of old oceans (while the eroded areas of old continents become new oceans).

Each stage automatically entails the next. The weight of accumulating sediments generates enough heat to consolidate, and then to uplift, the strata; the steep topography of uplift must then, perforce, erode as waves and rivers do their work. Time's cycle rules the world machine of erosion, deposition, consolidation, and uplift; continents and oceans change places in a slow choreography that can never end, or even age, so long as higher powers maintain the current order of nature's laws. Deep time becomes a simple deduction from the operation of the world machine.

The Hutton of Legend

Charles Lyell's self-serving rewrite of geology's history (see Chapter 4) demanded a certain type of hero, and Hutton best fitted the requirements. Simple chauvinism decreed a British character, and Hutton prevailed (even though nearly half his *Theory of the Earth* presents long, untranslated quotations from French sources). Hutton was never considered a major figure by continental geologists. I don't think that he even had much influence upon the great flowering and professionalization of British geology following the founding of the Geological Society of London in 1807. For this first generation devoted its attention to the very kind of historical inquiry that Hutton eschewed (see the last section of this chapter). Hutton's paramountcy fulfilled a later need.

Lyell's construction of history portrayed the emergence of scientific geology as the victory of uniformitarianism over the previous torpor of fruitless speculation based on untestable catastrophes and other fanciful proposals that explained the past by causes no longer affecting the earth. Lyell's vision demanded a hero as empiricist— a man willing to do his patient dog work in the field, and to build proper theories as inductions from observed phenomena. Hutton was pressed into service in one of the most flagrant mischaracterizations ever perpetrated by the heroic tradition in the history of science. Hutton came to embody the mystique of fieldwork against forces of reaction.

In this standard myth, Hutton discovered deep time because he formulated the cardinal principle of empirical geology and used it to draw two central conclusions from his fieldwork. We are told that Hutton devised the principle of uniformitarianism, loosely translated in textbook catechism as "the present is the key to the past." Using this guide, Hutton then observed, *first,* that *granite* must be an intrusive rock, not a sediment (therefore a reflection of uplifting powers, not a product of decay); and, second, that *unconformities* provide direct evidence for *multiple* cycles of uplift and erosion. Hutton used these two crucial observations as the basis for inducing a cyclical theory of the world machine from field evidence.

The Huttonian legend did not begin right away. Lyell praised him highly enough, but more as a man who tried to extend Newton's program from space to time, than as a great empiricist (1830, I, 61–63). In a private letter (in K. M. Lyell, 1881, II, 48), Lyell commented that Hutton's system showed no great advance beyond Hooke or Steno.

The elevation of Hutton achieved its canonical form in the same work that classified Burnet among the villains and presented the empiricist myth in its most influential form—Sir Archibald Geikie's *The Founders of Geology* (1897). Geikie's Hutton is a paragon of objectivity, a cardboard ideal. "In the whole of Hutton's doctrine, he vigorously guarded himself against the admission of any principle which could not be founded on observation. He made no assumptions. Every step in his deductions was based upon actual fact, and the facts were so arranged as to yield naturally and inevitably the conclusions which he drew from them" (1905 ed., 314–315). Bowing to the primal mystique of geology, Geikie identified the source of these rigorous observations in fieldwork: "He went far afield in search of facts . . . He made journeys into different parts of Scotland . . . He extended his excursions likewise into England and Wales. For about 30 years, he never ceased to study the natural history of the globe" (1905, 288). Geikie then labeled the theory of his fellow Scotsman as "a coherent system by which the earth became, as it were, her own interpreter" (1905, 305).

Geikie's mythical Hutton has been firmly entrenched in geological textbooks ever since. Our students are still introduced to him as the first real empiricist in geology and, as such, the founder of our science: "The first to break formally with religion-shrouded tradition was James Hutton," proclaims the CRM textbook *Geology Today* (1973). Leet and Judson (1971, 2), for many years the best selling of all texts, stated baldly: "Modern geology was born in 1785 when James Hutton . . . formulated the principle that the same physical processes that are operating in the present also operated in the past." Using a scatological metaphor from the labors of Hercules, Marvin (1973, 35) wrote: "He made it his task to clear the geological Augean Stables of the encrusted catastrophist doctrine of over one thousand years."

Following Geikie's lead, the texts then identify Hutton's great insight with his fieldwork. Bradley (1928, 364) wrote: "Throughout Hutton's 'Theory' the inductive method of reasoning alone is used. He made the earth tell its own story." Seyfert and Sirkin, in another leading introductory text (1973, 6), attribute all Hutton's successes to his fieldwork, all his failures to his writing: "Even though Hutton's ideas were backed by careful field observations, his paper was written in such a difficult style that it was not widely read."

But the most forceful retelling of Hutton's myth transcends the one-liners of traditional texts. John McPhee, for worthy reasons of his own (a generally romantic view of nature, as I read him, and a commitment to preserve natural beauty in an age of unparalleled danger), has adopted Hutton to convey the mystique of fieldwork as both science and aesthetics. In *Basin and Range* (1980), McPhee explores the two great revolutions of geology—deep time and ceaseless motion (as embodied in plate tectonics). Since he followed geologists in the field and lived the second revolution, he has rendered plate tectonics with acute perception (and beautiful prose). But as he relied upon standard histories for the first revolution, he has given the Huttonian myth its most literate retelling since Geikie's invention.

McPhee sets Hutton and his opponents as precursors of a modern tension in geology (a dichotomy with long tendrils, for it also evokes such basic contrasts as romantic and mechanical approaches, or holistic and analytic procedures)—the flashiness of complex laboratory equipment, often operated by people with strong mathematical skills but little knowledge of rocks, versus "old-fashioned" field observation. Abraham Gottlob Werner, Hutton's chief opponent in the debate about granite, becomes a prototype for the threat of soulless laboratories, with a boost from Geikie himself:

> Some contemporary geologists discern in Werner the lineal antecedence of what has come to be known as black-box geology—people in white coats spending summer days in basements watching million-dollar consoles that flash like northern lights—for Werner's "first sketch of a classification of rocks shows by its meagerness how slender at that time was his practical acquaintance with rocks in the field." The words are Sir Archibald Geikie's . . . an accomplished geologist who seems to have dipped in ink the sharp end of his hammer.   (94)

(I find this analogy particularly revealing because so inapt. Werner was a lecturer and mining engineer, not a lab man in an age that, in any case, boasted little in the way of fancy equipment. This false comparison can only record the desire of some geologists to equate what they don't like today with something scorned from the past.)

Hutton, by contrast, fashions modern geology by observing nature patiently and directly. The cyclical theory of the world machine is slowly "discerned" (as McPhee writes) in the rocks:

> Wherever he had been, he had found himself drawn to riverbeds and cutbanks, ditches and borrow pits, coastal outcrops and upland cliffs; and if he saw black shining cherts in the white

chalks of Norfolk, fossil clams in the Cheviot hills, he wondered why they were there. He had become preoccupied with the operations of the earth, and he was beginning to discern a gradual and repetitive process measured out in dynamic cycles. (95–96)

Hutton Disproves His Legend

The traditional argument that Hutton induced his cyclical theory of the world machine from field observations, particularly on granite and unconformities, becomes even harder to understand when we recognize that Hutton's own record clearly belies his legend *prima facie*.

Simple chronology is evidence enough. Hutton presented his theory of the earth before the Royal Society of Edinburgh on March 7 and April 4, 1785, and published an abstract, describing the theory essentially in its final form, later that year. The first full version appeared in 1788, in volume 1 of the *Transactions of the Royal Society of Edinburgh*, followed in 1795 by the massive (and traditionally unreadable) two-volume *Theory of the Earth with Proofs and Illustrations*.

Hutton saw his first unconformity in 1787 at Loch Ranza, followed later that year by an example in the Tweed Basin, the subject for John Clerk of Eldin's drawing (Figure 3.1). In 1788, Hutton found his most famous unconformity at Siccar Point, took his friends to see it by boat, and inspired Playfair's awe at "the abyss of time."

When he presented his theory in 1785, Hutton had observed granite at only one uninformative location in the field. That summer, he visited several better sites, including the outcrop at Glen Tilt where he made the key observation (Figure 3.2) of granitic veins intruding the local schists as a thicket of fine fingers. (If granite were a sediment, it could not have forced its way into a thousand nooks and cracks of overlying schist. The granite, Hutton concluded, must have intruded in molten form from below. It must be younger than the schist, and a force of later uplift, not older and a sign of the earth's original construction.)

Thus, as G. L. Davies argued in his masterful dissection of the empiricist myth (1969), Hutton developed his theory in its final

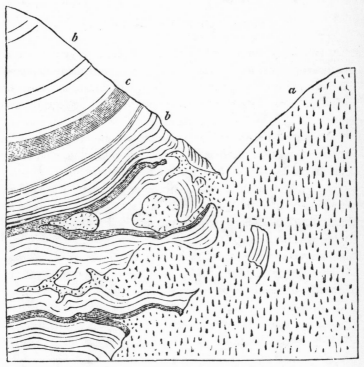

Junction of granite and limestone in Glen Tilt.

a, Granite.   *b*, Limestone.   *c*, Blue argillaceous schist.

Figure 3.2

A figure from the first edition of Charles Lyell's *Principles of Geology* (1830) showing the famous locality where Hutton confirmed the igneous nature of granite by finding a multitude of granitic fingers intruding into older sediments.

form before he had ever seen an unconformity, and when he had observed granite in only one inconclusive outcrop.

We might still support a weaker version of the empiricist myth if Hutton himself had espoused the mystique of fieldwork, and had attempted later to hide the *a priori* character of his theory by fudging the derivative character of his crucial observations. At least the ideal would remain intact.

Even this version fails before Hutton's own candor. He presents his theory—with pride—as derived by reason from key premises that have no standing in modern science (see next section). He then discusses his observations as subsequent confirmations of these ideas. His statement about granite could not be clearer or more concise: "I just saw it, and no more, at Petershead and Aberdeen, but that was all the granite I had ever seen when I wrote my Theory of the Earth [1788 version]. I have, since that time, seen it in different places; because I went on purpose to examine it" (1795, I, 214). As for unconformities, Hutton proclaims their derivative status in the chapter title for their discussion: "The theory confirmed from observations made on purpose to elucidate the subject" (1795, I, 453).

In fact, Hutton's work suffered gravely in reputation when a strong empiricist tradition did arise within geology early in the nineteenth century. Hutton's near contemporaries ranked him among the antiquated system-builders of a speculative age. Cuvier granted Hutton but a paragraph in his preliminary discourse of 1812, listing him second among six recent system-builders. Cuvier presented these six men as superior to purely speculative predecessors in their devotion to natural causes, but still in the armchair tradition, and mutually incompatible because such an inadequate methodology cannot attain consensus.[2]

---

2. Cuvier's crisp epitome of Hutton's cyclic world machine is worth repeating: "Les matériaux des montagnes sont sans cesse dégradés et entraînés par les rivières, pour aller au fond des mers se faire échauffer sous une énorme pression, et former des couches que la chaleur qui les durcit relèvera un jour avec violence."

In 1817, *Blackwood's Magazine* echoed the new empirical tradition and placed Hutton beyond the pale: "Had he studied nature, and then theorized, his genius would in all probability, have illustrated many difficult points; but it is obvious, from his own works, that he has frequently reversed this order of proceeding." Davies's judgment (1969, 178) is harsh, but not, I think, exaggerated or misplaced: "Mistitled, lacking in form, drowned in words, deficient in field evidence, and shrouded in an overall obscurity ... many of those who knew of the theory only through Hutton's expositions must have dismissed it as the worthless and indigestible fantasy of a somewhat outdated arm-chair geologist."

Hutton, in short, never misrepresented his intent. He viewed the earth as a body with a purpose. This purpose imposed requirements upon any rational theory—"things which must necessarily be comprehended in the theory of the earth, if we are to give stability to it as a world sustaining plants and animals" (1795, I, 281). Not only did Hutton deduce the necessity of a restorative force (the basis of cyclicity); he also stated repeatedly that his concept of a proper, purposeful universe would collapse if such a force could not be discovered.

When we finally discard the empiricist myth that turned Hutton into his opposite, we can properly seek the discovery of deep time in those *a priori* concepts that Hutton viewed as the rational basis for his or any theory of the earth. He did not find deep time or cyclicity in rocks. We can understand the role of time's metaphors in Hutton's geology—the "bottom line" of this chapter and book—only when we direct our search toward Hutton's systematic thinking rather than his field explorations.

The Sources of Necessary Cyclicity

Hutton is nothing if not consistent in the cycling of repeated themes through his thousand pages of text. Among these reiterations, no statement appears with more force or frequency than his insistence that any proper theory of the earth must explain its dual status—as a mechanism maintained by physical processes, and as an object constructed for a definite purpose. The opening paragraph of his first treatise calls the earth "a machine of a peculiar construction by which it is adapted to a certain end" (1788, 209). In 1795, he continues to unite means (or mechanisms) with ends (or purposes): "The theory of the earth shall be considered as the philosophy or physical knowledge of this world, that is to say, a general view of the means by which the end or purpose is attained; nothing can be properly esteemed such a theory unless it lead, in some degree, to the forming of that general view of things" (I, 270).

By mechanism, Hutton understands the cycling world machine itself. For the purpose of this cycling, he advances an unswerving conviction that we might brand as crass hubris today, but that seemed self-evidently true in his age. The earth was constructed as a stable abode for life, in particular for human domination. Again uniting means and ends, Hutton speaks of "this mechanism of the globe, by which it is adapted to the purpose of being a habitable world" (1788, 211). Extending the argument to human life, he writes of "a world contrived in consummate wisdom for the growth and habitation of a great diversity of plants and animals; and a world peculiarly adapted to the purpose of man, who inhabits all its climates, who measures its extent, and determines its productions at his pleasure" (1788, 294–295).

No notion is more alien to modern science than Hutton's insistence—as a pivot of his entire system, not peripheral verbiage—that physical objects have purposes shaped in human terms. Aristotle insisted that phenomena have at least four distinct kinds of causes—material for their substance, efficient for the pusher or builder, formal for the blueprint, and final for the purpose. A house, in the ancient parable, finds its material cause in stones or bricks, efficient in masons and carpenters, formal in architectural sketches (which "make" nothing in any direct sense, but are surely a *sine qua non* of any particular design), and final in human desires, for the house would not be built unless someone wanted to live in it. We have, today, pretty much restricted our definition of cause to the pushers and shovers of efficient causation. We would still allow that houses can't be built without substances and plans, but we no longer refer to these material and formal aspects as causes. We have, however, explicitly abandoned the idea of final cause for inanimate objects—and this rejection ranks as, perhaps, the major change in scientific methodology between Hutton's time and ours. We cannot understand Hutton until we recover his concept of final cause as a centerpiece of explanation.

We still speak of final cause for objects built with evident goals by human consciousness. We also permit a vernacular meaning of purpose in describing the adaptations of organisms, though a marlin does not strive consciously for hydrodynamic efficiency. But we have rigidly abjured any idea of final cause for inanimate objects, and we judge nothing more amusing or antiquated than previous attributions of purpose made in human terms—the moon shines so that we do not stumble at night, or oranges grow in sections so that we may easily divide them. We smile at Aristotle when he proposes both an efficient and a final cause for earthquakes: "It thunders both because there must be a hissing and roaring as the [earth's internal] fire is extinguished, and also (as the Pythagoreans hold) to threaten the souls in Tartarus and make them fear" (*Organon, Posterior Analytics*, 94b, l.34). Our laughter surely represents an inappropriate approach to history, but it does express the profundity of our change in attitude.

We cannot grasp the basis of Hutton's cyclical theory until we understand his commitment to final cause as a necessary ingredient of any explanation. In introducing his first treatise, Hutton says of the earth: "We perceive a fabric erected in wisdom, to obtain a purpose worthy of the power that is apparent in the production of it" (1788, 209). He advocates as a general methodology, the simultaneous search for both efficient and final causes expressed in human terms:

> Nothing can be admitted as a theory of the earth which does not, in a satisfactory manner, give the efficient causes for all these effects ... But this is not all. We live in a world where order every where prevails; and where final causes are as well known, at least, as those which are efficient. The muscles, for example, by which I move my fingers when I write are no more the efficient cause of that motion, than this motion is the final cause for which the muscles have been made. Thus, the circulation of the blood is the efficient cause of life; but, life is the final cause, not only for the circulation of the blood but for the revolution of the globe ... Therefore the explanation, which is given of the different phenomena of the earth, must be consistent with the actual constitution of this earth as a living world, that is, a world maintaining a system of living animals and plants. (1795, II, 545–546)

Hutton presents his theory as the *a priori* solution to a problem in final causation, not as an induction from field evidence. We might choose to disregard Hutton's own insistent claim, and argue that no one could really base so much on what seems so nonsensical today. But the intellectual bankruptcy of such an attitude should be self-evident.

Hutton states explicitly, at the outset of his first treatise (1788, 214–215), and throughout all his writing, that his theory is an argument made *a priori*, and logically necessary to resolve a paradox in final cause. Why not take him at face value? We may call this problem the "paradox of the soil." Hutton had spent most of his early life, before retiring to intellectual circles in Edinburgh, as a committed and successful gentleman farmer, studying and using the latest methods of husbandry. He had thought long and hard about soil, the substrate of agriculture, and all life. Soil must be rich and constant to fulfill the earth's final cause as an abode for life.

Soil, generated from eroding rocks, is a product of destructive forces: "For this great purpose of the world, the solid structure of this earth must be sacrificed; for, the fertility of our soil depends upon the loose and incoherent state of its materials" (1795, II, 89). But if the destruction of land continue unabated, continents will eventually wash into the sea: "The heights of our land are thus levelled with the shores; our fertile plains are formed from the ruins of the mountains" (1788, 215). The process that sustains life will eventually destroy it: "The washing away of the matter of this earth into the sea would put a period to the existence of that system which forms the admirable constitution of this living world" (1795, I, 550).

Efficient causes on a benevolent earth cannot undermine the final causes of stability for human life. Yet the soil undoubtedly arises by destruction. Hutton therefore argues that a restorative force must exist to rebuild the continents. Moreover, if the source of uplift can be rendered as a consequence of prior destruction, then our earth embodies the simplest and most harmonious of possible systems—not two independent forces of breaking and making locked in delicate balance, but a single cycle automatically sustaining a steady state of benevolence. If erosion not only makes soil but also deposits strata for continents of the next cycle, the paradox of the soil can be resolved with elegance: "But, if the origin of this earth is founded in the sea, the matter which is washed away from our land is only proceeding in the order of the system; and thus no change would be made in the general system of this world, although this particular earth, which we possess at present, should in the course of nature disappear" (1795, I, 550).

Hutton could not have stated more clearly that he deduced the necessary existence of uplifting forces as a required solution to the paradox of the soil—a dilemma in final cause. Deep time, inherent in the resulting cyclicity, belongs to the logical structure of his *a priori* argument. In my favorite passage, Hutton tells us why final cause requires restoration and cyclicity:

> This is the view in which we are now to examine the globe: to see if there be, in the constitution of this world, a reproductive operation by which a ruined constitution may be again repaired, and a duration or stability thus procured to the machine, considered as a world sustaining plants and animals.
>
> If no such reproductive power, or reforming operation, after due enquiry, is to be found in the constitution of this world, we should have reason to conclude, that the system of this earth has either been intentionally made imperfect, or has not been the work of infinite power and wisdom.   (1788, 216)

The *a priori* character of cyclicity and deep time inheres just as strongly in Hutton's attitude toward mechanisms, or efficient causes. The earth requires a restorative force to fulfill its purpose as an abode for life, but how does uplift occur? I discussed the mechanics of Hutton's cycle earlier in this chapter, but what led him to a theory of this kind, or to a notion of self-sustaining cycling at all, since he didn't just see cycles in the field?

The sources of Hutton's world machine are complex, but one influence stands out in his writing. The light of Newton's triumph continued to shine brightly, and the union of other disciplines with the majesty of his vision remained a dream of science at its best. Hutton yearned to read time as Newton had reconstructed space. If the apparent messiness of complex history could be ordered as a stately cycle of strictly repeating events, then the making and unmaking of continents might become as lawlike as the revolution of planets.

Hutton's world machine is Newton's cosmos read as repeating order through time. The discovery of a restorative force, Hutton argues, fixes the analogy and guarantees time without limit to the earth under its current management of natural law: "When he finds that there are means wisely provided for the renovation of this necessarily decaying part, as well as that of every other, he then, with pleasure, contemplates this manifestation of design, and thus connects the mineral system of this earth with that by which the heavenly bodies are made to move perpetually in their orbits" (1795, I, 276). Hutton also invokes a cosmic analogy as a guarantee of deep time in the sentence just preceding his famous closing dictum, "no vestige of a beginning,—no prospect of an end": "For having, in the natural history of this earth, seen a succession of worlds, we may from this conclude that there is a system in nature; in like manner as, from seeing revolutions of the planets, it is concluded, that there is a system by which they are intended to continue those revolutions" (1788, 304). Similarly, Playfair connects deep time with planetary motion:

> The geological system of Dr Hutton resembles, in many respects, that which appears to preside over the heavenly motions . . . In both, a provision is made for duration of unlimited extent, and the lapse of time has no effect to wear out or destroy a machine, constructed with so much wisdom. Where the movements are all so perfect, their beginning and end must be alike invisible. (1802, 440)

In summary, I have traced the *a priori* character of cyclicity and deep time in Hutton's thought by analyzing his views on the nature of final and efficient causes for the earth. For final cause, he resolved the paradox of the soil by insisting that uplift must restore topography eroded to permit life and agriculture. For efficient cause, he devised a world machine that arranged all historical complexity as a cycle of repeating events as regular as the revolution of planets in Newton's system. In both cases, deep time is the essential ingredient of unbounded cycles, established by logical necessity prior to confirmation in the field. In other words—and I may now summarize the entire chapter in a phrase—time's cycle forms the core of Hutton's vision for a rational theory of the earth. *Hutton developed his theory by imposing upon the earth the most rigid and uncompromising version of time's cycle ever developed by a geologist.*

We may appreciate Hutton's audacity, and his success in breaking the bonds of time by a strategy that exalted one central metaphor and excluded the other. Hutton's theory of the earth is time's cycle triumphant; but can his total rejection of time's arrow pass without rueful consequence?

Hutton's Paradox: Or, Why the Discoverer of Deep Time Denied History

The Pure Time's Cycle Theorist

If moments have no distinction, then they have no interest.

I propose this aphorism as a description of Hutton's paradox, or the problematical situation that pure versions of time's cycle impose upon history. We saw in Chapter 2 how Burnet insisted, so acutely, that any strict reading of time's cycle would rob him of his subject. I wish to argue that Hutton's approach implied such an attitude toward history, and that he at least had the gumption and consistency to follow his argument to its logical end, and thereby to deny history itself. Such a claim will appear, particularly to most geologists, as absurd. After all, Hutton discovered deep time, didn't he? How could the architect of a proper matrix for history then turn upon his own implication and deny it? Yet Hutton proceeded just this way—and we have lost the resulting paradox, both because we know Hutton from Playfair's different translation (a less rigid version that permitted history), and because we have not understood the centrality and power of time's cycle in Hutton's argument.

The paradox is both logical and psychological. As a bare minimum, history demands a sequence of distinctive events (other issues like directionality and rates of transition are subjects of endless debate and fascination, but not *sine quibus non*). Under the metaphor of time's cycle in its pure form, nothing can be distinctive because everything comes round again—and no event, by itself, can tell us where we are, for nothing anchors us to any particular *point* in time, but only (at most) to a particular stage of a repeating cycle.

The psychological argument is a simple matter of interest: why be concerned with the apparently distinctive details of any geological event if it possesses no individuality, but represents one of a potentially endless class? We discuss with relish the idiosyncrasies of Bill the cat, but who ever talks about Joe the silica tetrahedron?

The clearest evidence of Hutton's adherence to a rigid version of time's cycle lies in his explicit denials of history and his avoidance of all metaphors involving sequence and direction. Hutton tells us that the earth's cycles lead nowhere; he does not permit Burnet's resolution of cycles advancing as they turn—the model of a large disc rolling down a railroad track. The last cycle was no different from nature's current course, for it witnessed "an earth equally perfect with the present, and an earth equally productive of growing plants and living animals" (1788, 297). Change is a continuous backing and forthing, never a permanent alteration in any direction: "At all times there is a terraqueous globe, for the use of plants and animals; at all times there is upon the surface of the earth dry land and moving water, although the particular shape and situation of those things fluctuate, and are not permanent as are the laws of nature" (1795, I, 378–379).[3]

Most revealing are Hutton's methodological statements about the role of those quintessential data of history—sequences of events in time. He does not view them, in any sense, as components of narrative interesting in themselves, but only as data to use in establishing general theories of timeless systems. Again, making his favorite Newtonian analogy, Hutton writes:

> In order to understand the system of the heavens, it is necessary to connect together periods of measured time, and the distinguished places of revolving bodies. It is thus that system may be observed, or wisdom, in the proper adapting of powers to an intention. In like manner, we cannot understand the system of

3. Note Hutton's choice of words—"fluctuate," with its implication of motion back and forth about a constant average, not "change," which might imply an element of directionality.

the globe, without seeing that progress of things which is brought about in time.   (1788, 296)

We saw in Chapter 2 how Burnet expressed his intricate melding of arrows and cycles with a corresponding mix of metaphors appropriate for both. Hutton's metaphors, by contrast, are striking in their exclusivity. He invokes all the standard-bearers of balance and repetition in our culture, and no symbols whatever for direction or progress. We have already explored his primary comparison—a cycling earth with revolving planets of Newton's cosmos. Hutton's ahistoric world is a dynamic balance of opposing forces, not a passive stability, and his metaphors record the dynamic steady state of his cycling system. Thus, planets stay in their orbits because a linear force that would propel them ever farther away balances a gravitational force that would pull them into the sun (1788, 212), just as the stability of time's cycle balances destruction and renovation. Planetary motions also establish a set of shorter cycles forming abundant material for metaphor: days, seasons, and all the repetitions described by Hutton under a general rubric of the earth's fundamental "economy," or balance: "With such wisdom has nature ordered things in the œconomy of this world, that the destruction of one continent is not brought about without the renovation of the earth in the production of another" (1788, 294).

Most revealing are Hutton's uses of the human body in metaphor, for our lives, unlike revolving planets, offer abundant material for *either* arrow or cycle metaphors. Yet Hutton avoids the obvious directional themes—growth, learning, development on the one hand; decline, aging, and death on the other—so avidly embraced by Burnet and other exponents of time's arrow. Instead, he invokes only those aspects of life that maintain our bodies, or our populations over generations, in steady state. The circulation of our blood resembles the hydrologic cycle that erodes continents (Hutton's own doctoral dissertation as a medical student at Leiden, had treated the circulation of blood): "All the surface of this earth is formed according to a regular system of heights and hollows, hills and valleys, rivulets and rivers, and these rivers return the waters of the atmosphere into the general mass, in like manner as the blood, returning to the heart, is conducted in the veins" (1795, II, 533). But the earth's renovation of its eroded topography recalls the processes of growth, feeding, and healing that restore an animal's body: "We are thus led to see a circulation in the matter of this globe, as a system of beautiful œconomy in the works of nature. This earth, like the body of an animal, is wasted at the same time that it is repaired. It has a state of growth and augmentation; it has another state, which is that of diminution and decay" (1795, II, 562).

Finally, these cycles of erosion and renovation proceed hand in hand, just as human births balance deaths to maintain a stability of population through the ages. Consider, as a summary of Hutton's metaphors, this commingling of his two favorite sources, planets and bodies:

> Why refuse to see, in this constitution of things, that wisdom of contrivance, that beautiful provision, which is so evident, whether we look up into the great expanse of boundless space, where luminous bodies without number are placed, and where, in all probability, still more numerous bodies are perpetually moving and illuminated for some great end; or whether we turn our prospect towards ourselves, and see the exquisite mechanism and active powers of things, growing from a state apparently of non-existence, decaying from their state of natural perfection, and renovating their existence in a succession of similar beings to which we see no end. (1795, II, 468–469)

Perfection and the Denial of History

I have traced Hutton's direct and metaphorical statements disavowing any interest in history; I shall document shortly his peculiar treatment of history's primary data in geology (fossils and strata).

But what, beyond the obvious mechanics of cycling in his world machine, made Hutton so singularly uninterested in narrative?

Many great arguments in the history of human thought have a kind of relentless, intrinsic logic that grants them a universality transcending time or subject. In such cases, we can, with due respect to differences of age and culture, make comparisons that illuminate the generality of an argument by its very congruence through such different circumstances. Hutton's primary reason for denying history falls within an argument of this scope.

A quiet intellectual struggle has pervaded evolutionary biology ever since Darwin developed the theory of natural selection—a tension between optimal design and history. Some strict Darwinians have located the beauty of natural selection in its ability to produce optimal forms as adaptations—and they may wax lyrical about the aerodynamic perfection of a bird's wing, or the ideal camouflage of a butterfly mimicking a dead leaf. Others have viewed such pan-selectionism as a subtle perversion of the subject itself. Evolution is the conviction that organisms developed their current forms by an extended history of continual transformation, and that ties of genealogy bind all living things into one nexus. Panselectionism is a denial of history, for perfection covers the tracks of time. A perfect wing may have evolved to its current state, but it may have been created just as we find it. We simply cannot tell, if perfection be our only evidence. As Darwin himself understood so well, the primary proofs of evolution are oddities and imperfections that must record pathways of historical descent—the panda's thumb and the flamingo's smile (Gould, 1983, 1985) of my book titles (chosen to illustrate this paramount principle of history).

This principle of imperfection is a general argument for history, not a tool of evolutionary biologists alone. All historical scientists use it, as Burnet did in likening a ruined earth to the destruction of Solomon's temple as comparable evidence for history in non-optimal structures; as linguists must in detecting history when current usage does not match etymology (consider the bucolic basis of "broadcasting," sowing seed, or an "egregious" object as outside the flock, *ex grege*).

In reverse, then, perfection becomes an argument against history—a denial, at least, of its importance, sometimes of its very existence. The historical antecedents of any optimal state become irrelevant either because the system now stands in perfect, timeless balance or, in the stronger version, because different stages never existed, and wisdom made perfection from the start.

In this sense, Hutton's strongest argument against history flows necessarily from his passionate conviction about the perfection of his world machine; no other theme so pervades his works, or so underlies his insistent comparison of earthly time with celestial machinery. How can a historical narrative of change be relevant to a perfectly working machine fulfilling its ordained purpose from its inception?

The basic components of narrative are, to Hutton, the very definitions of imperfection: the punctuation of time by peculiar and random large-scale events; and, particularly, a lack of cyclicity defined as any kind of directional change—for if things improve in time, then the world machine was not made perfect, and if they decline, then the earth is not perfect now. In comparing himself with Burnet and other exponents of geological history as decline from an original perfection, Hutton defends optimality as the greatest virtue of his world machine: "In discovering the nature and constitution of this earth . . . there is no occasion for having recourse to any unnatural supposition of evil, to any destructive accident in nature, or to the agency of any preternatural cause, in explaining that which actually appears" (1788, 285).

It would be irrational, Hutton argues, to defend the optimal constancy of ecological balance between plants and animals (a proven fact in Hutton's view), and then to argue that their earthly substrate is wasting away to destruction: "To acknowledge the perfection of those systems of plants and animals perpetuating their species, and to suppose the system of this earth on which they must depend, to be imperfect, and in time to perish, would be to reason inconsistently or absurdly" (1795, I, 285).

In a striking example of their differences, Hutton echoes Burnet's words (see page 42) in stating that people have a deep and unquenchable desire to understand sequences of events in time: "Man is not satisfied, like the brute, in seeing things which are; he seeks to know how things have been, and what they are to be" (1788, 286). We might almost think that Hutton is cranking up for a defense of history. But, whereas Burnet uses this preamble to glorify the intrinsic fascination of narrative, Hutton takes the opposite line dictated by his allegiance to time's cycle: we want to understand what happened in time only so that we may infer the cycling, timeless system of change, and thereby grasp the perfection of nature's works. Continuing directly from the last quotation: "It is with pleasure that he observes order and regularity in the works of nature, instead of being disgusted with disorder and confusion; and he is made happy from the appearance of wisdom and benevolence in the design, instead of being left to suspect in the Author of nature, any of that imperfection which he finds in himself."

Divorcing History from Its Own Best Data

The classical data of historical geology are fossils and strata. Obviously, we cannot charge Hutton with inattention to principles that were codified after his death. In particular, Hutton's contemporaries had not resolved the issue of extinction and fossil sequences. Lamarck and others were still arguing that species could not die, and Cuvier's proof of extinction, with its guarantee that history might be calibrated by the distinctive life-spans of fossil groups, followed Hutton's death by a decade or more. But basic stratigraphic principles of superposition and correlation had been developed. Maps and sections, however rudimentary, were being published—though not by Hutton. A rude system of stratigraphic nomenclature had been developed to order events in time—the "primary" cores of mountains, the hard "secondary" strata deposited against them, and the still younger, loosely consolidated "tertiary" deposits (the last name still surviving, in more prestigious upper case, as a period of the Cenozoic Era).

Hutton used the data of fossils and strata as primary empirical supports for his system, but he never invoked them as signs of history. Since we cannot attribute this failure entirely to ignorance of principles unknown in his day, Hutton's curiously limited use of these data does reflect his resolutely ahistorical perspective.

Hutton on fossils

Although paleontological data provided crucial information toward validating several parts of the world machine, we find in Hutton's writing not a shred of a suggestion that fossils might record a vector of historical change, or even distinctness of moments in time. Fossils, to Hutton, are immanent properties of time's cycle.

The encasing of marine fossils in continental strata illustrates two essential parts of the world machine: first, their incorporation into hard strata proves that piles of sediment can be consolidated to rock by heat and pressure; second, their present status as parts of elevated continents demonstrates that consolidated sediments are then uplifted by restorative forces. "In all the regions of the globe, immense masses are found, which, though at present in the most solid state, appear to have been formed by the collection of the calcareous exuviae of marine animals" (1788, 219).

But how can we know that these marine sediments formed from eroded materials of continents in a former cycle? Here, Hutton invokes petrified wood and other plant fossils (1788, 290–292) as direct proof for vanished continents. In other words, both examples use fossils only as ecological signs in judging sources and places for deposition of sediments, not as historical evidence for distinctive changes in time. Hutton denies that any change at all has accompanied life's passage through time's cycle:

> In order to be convinced of that truth, we have but to examine the strata of our earth, in which we find the remains of animals. In this examination, we not only discover every genus of animal which at present exists in the sea, but probably every species, and perhaps some species with which at present we are not acquainted. There are, indeed, varieties in those species, compared with the present animals which we examine, but no greater varieties than may perhaps be found among the same species in the different quarters of the globe. (1788, 290)

The last sentence of this quotation is particularly revealing. Hutton argues that if we find, within fossil species, varieties unknown among living forms, these varieties are probably not distinctively ancient, but merely as yet undiscovered among living creatures. This choice between competing hypotheses, made by preference and without evidence, shows that alternatives to Hutton's belief in constancy were debated in his time—and that his denial of history is an active preference, not a simple citation of contemporary consensus.

In the one passage where Hutton dares not deny distinctive difference in time, he manages to bypass the subject completely, using another aspect of the tale to support time's cycle. Hutton does not argue that human life has pervaded time, but admits the scriptural tradition of recent origin. He simply acknowledges our late appearance in a sentence, then immediately moves on to extolling other fossils as indicators of deep time:

> The Mosaic history places this beginning of man at no great distance; and there has not been found, in natural history, any document by which a high antiquity might be attributed to the human race. But this is not the case with regard to the inferior species of animals, particularly those which inhabit the ocean and its shores. We find in natural history monuments which prove that those animals had long existed.   (1788, 217)

Hutton on strata

Reading Hutton's chapter on unconformities (1795, I, ch. 6) must be an unnerving experience for any geologist (though few have ever dipped into the original text). Hutton does everything that any good field geologist would do: he maps, he traces beds, he studies sequences in superposition. He talks about primary and secondary strata as older and younger, using their difference in time (and their separation by an unconformity) as evidence of process. He presents his descriptions as historical sequences. Writing, for example, about lower and upper units separated by an unconformity: "Here we further learn, that the indurated and erected strata, after being broken and washed by the moving waters [during formation of the unconformity], had again been sunk below the sea, and had served as a bottom or basis on which to form a new structure of strata" (1795, I, 449).

Yet Hutton's interpretations are decidedly peculiar, when judged against long traditions of field study from his own day through ours. These historical data are never cited as narrative. Through the thousand pages of Hutton's treatise, we find not a single sentence that treats the different ages and properties of strata as interesting in themselves—as markers of distinction for particular times. Never even the most basic statement that at some particular time, some definite environment led to the deposition of this kind of rock in that specific place. We learn instead that recognizable, temporally ordered strata affirm a general theory of time's cycle and the world machine: "By thus admitting a primary and secondary in the formation of our land, the present theory will be confirmed in all its parts. For, nothing but those vicissitudes, in which the old is worn and destroyed, and new land formed to supply its place, can explain that order which is to be perceived in all the works of nature" (1795, I, 471–472).

The earlier treatise of 1788 is even more explicit in rejecting narrative. Hutton states that our gut-level interest in "the oldest" is undermined by time's cycle, for we recognize that the bottom of a stratigraphic pile is sediment derived from older continents, and so forth to a beginning without vestige:

We are now to take a very general view of nature, without descending into those particulars which so often occupy the speculations of naturalists, about the present state of things. We are not at present to enter into any discussion with regard to what are the primary and secondary mountains of the earth; we are not to consider what is the first, and what the last, in those things which now are seen.   (1788, 288)

Hutton and the methods of history

One cannot criticize a person for ignoring what he had no reason to consider. If Hutton had been a physicist who never worked with the data of history, my comments would be out of place. But Hutton not only used such data; he also showed deep understanding of methods for historical inference.

In studying Darwin, I have tried to show (Gould, 1982, 1986a) that the development of a general methodology for historical inference forms the coordinating theme for all his books. I have arranged his methods as a sequence of different strategies in the face of decreasing information. As I read Hutton, and became more impressed by his subtle understanding of historical inquiry, I found that he used all Darwin's methods. The finest illustration of Hutton's actively ahistorical focus lies in his masterful understanding of how history may be inferred, followed by explicit rejection of the subject for itself, and the marshaling of its data to establish a general theory that makes history uninteresting.

Consider two examples. In best cases, we know the process that produced past events and can observe its operation today. We extrapolate current rates through time to see if continued operation can yield the full extent of past phenomena. This is uniformitarianism in its pure form. The major stumbling block for this method lies in a popular perception that change so slow is no change at all; but deep time provides a matrix that converts the imperceptible to the mightily efficient. Hutton uses this argument to maintain that slow erosion by streams and waves will destroy the continents (as Darwin argued that natural selection of tiny changes extrapolates to major trends in evolution, or that worms, working slowly and unnoticed beneath our feet, will in time shape the topography of England):

> The object which I have in view, is to show, first, that the natural operations of the earth, continued in a sufficient space of time, would be adequate to the effects which we observe; and secondly, that it is necessary, in the system of the world, that these wasting operations of the land should be extremely slow. In that case, those different opinions would be reconciled in one which would

explain, at the same time, the apparent permanency of this surface on which we dwell, and the great changes that appear to have been already made. (1795, II, 467–468)

But we often have no direct data rooted in modern, observable processes. In such cases, we must gather a multiplicity of past results and try to order them as reasonable stages in the operation of a single historical process (as Darwin did in arguing that fringing reefs, barrier reefs, and atolls represent three stages in the subsidence of island platforms). Hutton uses this method to interpret sequences of deposition and distortion of strata by uplift and tilting:

All those strata of various materials, although originally uniform in their structure and appearance as a collection of stratified materials, have acquired appearances which are often difficult to reconcile with that of their original, and is only to be understood by an examination of a series in those objects, or that gradation which is sometimes to be perceived from the one extreme state to the other, that is from their natural to their most changed state. (1795, II, 51)

Hutton's one expression of history: a small irony

If history be narrative moving somewhere through unique stages, then we cannot find history in Hutton's world machine. Only once in his entire treatise do we get a whiff of change as directional progress. This concept appears nowhere in his science (or even in his own words), but only in the flowery and obligatory puff of introductory praise to the monarch who served as titular head and sponsor of the Royal Society of Edinburgh—ironically, since he was America's tormentor so negatively described by Mr. Jefferson in the Declaration of Independence, none other than George III, "a monarch who has distinguished his reign by the utility of his institutions for improving the elegant arts, as well as by the splendor and success of his undertakings to extend the knowledge of Nature" (from Buccleugh's introduction to volume 1 of the *Transactions*, where Hutton's 1788 treatise appeared).

Borges's Dilemma and Hutton's Motto

I designated as Borges's dilemma the incomprehensibility that true eternity imposes upon our understanding (see page 48). Hutton had to resolve this logical conundrum, since he believed so strongly that Newtonian science required a pure vision of time's cycle for the mechanics of earthly processes, and that no event could therefore gain distinction in history. Hutton avoided Borges's dilemma with a brilliant argument that doubled as incisive methodology about what science can and cannot do. He held that time's cycle governs the earth only while it operates under the regime of natural laws now in force. These laws prescribe the cycle of the world machine and therefore provide no insight about beginnings and ends. Logic demands both beginnings and ends, but ultimate origins lie outside the realm of science. Some higher power established the current regime of natural laws at an unknowable time in the distant past, and will terminate this reign at an undetermined moment in the future—but science cannot deal with such ultimates.

Thus, Hutton chose his most famous words with consummate care, though posterity has often misread him as an exponent of infinite time. We see "no *vestige* of a beginning"—but the earth had an inception now erased from geological evidence by the cycling of its products through so many subsequent worlds. We discern "no *prospect* of an end" because the current regime of natural law cannot undo our planet—but the earth will terminate, or change to a different status, when higher powers choose to abolish the current regime. With one stroke, Hutton both gained the benefit and avoided the dilemma of time's cycle in its pure form. He acquired the virtue (as he saw it) of a perfect, repeating system with no peculiarities of history to threaten the hegemony of a timeless set of causes; and he resolved Borges's dilemma by relegating beginnings and ends, the anchors that comprehension requires, to a realm outside science. As Playfair wrote in summary: "Thus he arrived at the new and sublime conclusion, which represents nature as having provided for a constant succession of land at the surface of the earth, according to a plan having no natural termination, but calculated to endure as long as those beneficent purposes, for which the whole is destined, shall continue to exist" (1805, 56–57).

Playfair: A Boswell with a Difference

This long exegesis of time's cycle and its meaning for Hutton has left one essential question unanswered. If I am right, and the Hutton of our textbooks is the Hutton of history turned on his head, why have we read him so wrongly, and with such consistency in error? How could we have taken such a brilliant man, driven by such a powerful vision of time's cycle imposed upon the earth to solve a problem in final causality, and reconstructed him as a modern empiricist, a field geologist dedicated only to efficient causes? Geikie may have perpetrated this myth, but how did he get away with it? People are not so stupid. Could they possibly have read Hutton, however blinded by expectation, and found Geikie's version within?

The answer must lie, in large part, with Hutton's legendary unreadability.[4] By long tradition, and by simple unavailability, geologists do not read Hutton himself. Nineteenth-century Britain was blessed with a number of fine scientists who were also superb literary stylists—Charles Lyell and T. H. Huxley in particular. But the best writer of all may have been John Playfair, professor of mathematics at Edinburgh, dedicated amateur geologist, and intimate friend of Hutton. After Hutton's death, Playfair decided to rescue his friend's ideas from their poor presentation by publishing

---

4. I have never found Hutton nearly so obtuse or infelicitous as tradition dictates. I will not defend the rambling thousand-page 1795 *Theory* with its endless quotations in French (one runs for forty-one pages), but I find the 1788 version reasonably crisp and concise, with occasional lines of literary brilliance. Still, the historical record speaks for itself. Lyell admitted that he had never managed to read it all. Even Kirwan, Hutton's dogged, almost frantic critic (he opposed Hutton in whole books), never read all of both volumes—for many pages of his personal copy are uncut (see Davies, 1969).

a shorter volume describing the Huttonian theory in clearer form. We know Hutton almost exclusively from Playfair's beautiful and successful exposition, *Illustrations of the Huttonian Theory of the Earth* (1802).

Tradition also dictates that Playfair simply translated his friend's ideas without alteration—so that, in the *Illustrations*, we really do read pure Hutton spruced up. In one sense, I do not deny this claim. The essence of Hutton's system receives an accurate and sensitive description in Playfair's writing. Time's cycle, in particular, appears in unvarnished form, with appropriate Newtonian analogies and incisive comparisons. I particularly treasure Playfair's contrast of Buffon's historical earth, declining to destruction by loss of heat, with Hutton's timeless cycles. Note Playfair's equation of time's cycle with rationality itself:

> Buffon represents the cooling of our planet, and its loss of heat, as a process continually advancing, and which has no limit, but the final extinction of life and motion over all the surface, and through all the interior, of the earth. The death of nature herself is the distant but gloomy object that terminates our view, and reminds us of the wild fictions of the Scandinavian mythology, according to which, annihilation is at last to extend its empire even to the gods. This dismal and unphilosophic vision was unworthy of the genius of Buffon, and wonderfully ill suited to the elegance and extent of his understanding. It forms a complete contrast to the theory of Dr Hutton, where nothing is to be seen beyond the continuation of the present order; where no latent seed of evil threatens final destruction to the whole; and where the movements are so perfect, that they can never terminate of themselves. This is surely a view of the world more suited to the dignity of Nature and the wisdom of its Author.   (485–486)

Yet, in another sense, I find a universe of difference between Hutton and Playfair—a distinction that has been missed because Hutton has not been understood as a theorist of time's cycle who denied history. These are the parts of Hutton's work that seem most unacceptable and archaic in the light of geology's later traditions. And these are the aspects of Hutton's thought that Playfair either soft-pedals or presents in altered light. Playfair subtly "modernized" his friend, and helped to set the basis of Hutton's legend by toning down his hostility to history.

In one important change, Playfair largely excises Hutton's commitment to final cause. He does not deny his friend's obsession with a style of science already becoming archaic. Playfair even acknowledges the primacy of final cause in Hutton's system: "He would have been less flattered, by being told of the ingenuity and originality of his theory, than of the addition which it had made to our knowledge of final causes" (1802, 122). But whereas final cause and purpose are relentless themes on every page of Hutton's theoretical discussion, Playfair hardly mentions the subject. I can find only two passages that discuss final cause explicitly (121–122 and 129), while hundreds of Playfair's pages recount the mechanics of Huttonian cycles in a world of efficient causation.

But Playfair's most striking change is an alteration of sense, not emphasis. In discussing field evidence, Playfair follows the primary tradition of geology from its inception, and does not portray Hutton's primary idiosyncrasy—his denial of history.[5] Playfair covers the same ground as Hutton, but his discussions of unconformities (for example) express the traditional interest of geologists in history for itself, whereas Hutton used historical events only to establish his cycling world machine, never to record the slightest concern for unique happenings in time.

Playfair discusses Hutton's unconformity (see Figure 3.1) as a sequence of distinctive occurrences in time. He argues that the picture displays evidence for three worlds in succession, and he discusses them from oldest to youngest. He notes that the bottom strata contain sand and gravel from the dissolution of a world still older—"the most ancient epocha, of which any memorial exists in

5. This distinction seems to me vitally important, yet I believe that all commentators have missed it. This difference has not been noted, I think, because Playfair's descriptions in the historical mode seem so obvious and "natural" that they have not been deemed odd or worthy of note as potentially distinct from Hutton—all because Hutton's ahistorical perspective has not been properly documented.

the records of the fossil kingdom" (123). Playfair clearly cares about old for old's sake, and he notes with pleasure that continents wasted to produce vertical strata below the unconformity represent a world "third in succession" (123) back from our present earth.

Continuing the historical sequence, Playfair discusses vertical strata below the unconformity and marvels at the vicissitudes of history. These rocks were broken and raised, lowered to receive sediments above the unconformity, then raised onto continents a second time—"so that they have twice visited the superior and twice the inferior regions" (123). They also represent the second world of this historical sequence. Playfair then moves on to the horizontal strata above the unconformity—the third world—and brings his narrative to the latest event of erosion, "the shaping of all the present inequalities of the surface" (124). Whereas Hutton disdained to record sequential events, Playfair orders all these stages into history. He concludes: "These phenomena, then, are all so many marks of the lapse of time, among which the principles of geology enable us to distinguish a certain order, so that we may know some of them to be more, and others to be less distant" (124–125).

Playfair's historical descriptions seem so simple, so innocent, so obvious. How could they mark a major departure? Yet you may read a thousand pages of Hutton's *Theory* and never find a phrase written in this mode. In short, Playfair won greater acceptability for Hutton by portraying his field evidence in the traditional, historical style that Hutton himself had consistently shunned. Even Hutton's Boswell could not follow his friend's rigorously ahistorical tastes, a predilection so contrary to our ordinary interest in the distinctive arrangement of things in time.

A Word in Conclusion and Prospect

Hutton "discovered" deep time by imposing his rigid view of time's cycle upon a complex earth. He did so, in part, to resolve a paradox in final cause—an issue that is no longer part of science. But his other motivation echoes a theme of outstanding relevance today. Hutton did not grasp the power, worth, and distinction of history. He followed a model of science that exalted simple systems, subject to experiment and prediction, over narrative and its irreducible uniquenesses. In so doing, he followed a tradition of ordering the sciences by status—from the hard and more "experimental" (physics and chemistry) to the soft and more "descriptive" (natural history and systematics). Geology resides in the middle of this false continuum, and has often tried to win prestige by aping the procedures of sciences with higher status, and ignoring its own distinctive data of history. This problem, born of low self-esteem, continues to our day. Hutton pursued a chimerical view of rigor by deference to Newton, and hoped to assimilate time to Newton's models for space. Today, this deference may be expressed in a fetish for quantification that leads psychologists to conceive intelligence as a single, measurable thing in the head, or biologists to classify organisms by computer without judging the different historical value of characters (the marsupial pouch as more informative than body length).

Charles Lyell recognized the link between Hutton and Newton, but he also noted an unhappy comparison—the triumph of cosmology versus the limited success of Hutton's world machine. He attributed this unflattering difference to the relative paucity of geological evidence, implying that diligence in collecting data might close the gap: "Hutton labored to give fixed principles to geology, as Newton had succeeded in doing to astronomy; but in the former science too little progress had been made towards furnishing the necessary data to enable any philosopher, however great his genius, to realize so noble a project" (1830, I, 61). I dedicate this book to a different view of this discrepancy: time's cycle cannot, in principle, encompass a complex history that bears irreducible signs of time's arrow. Hutton's rigidity is both a boon and a trap. It gave us deep time, but we lost history in the process. Any adequate account of the earth requires both.

AWFUL CHANGES.

MAN FOUND ONLY IN A FOSSIL STATE.——REAPPEARANCE OF ICHTHYOSAURI.

A Lecture.——" You will at once perceive," continued PROFESSOR ICHTHYOSAURUS, " that the skull before us belonged to some of the lower order of animals; the teeth are very insignificant, the power of the jaws trifling, and altogether it seems wonderful how the creature could have procured food.'

Figure 4.1

De la Beche's caricature of Charles Lyell as the future Professor Ichthyosaurus, as produced in the frontispiece of Frank Buckland's *Curiosities of Natural History*, and misinterpreted as a joke about his father, William Buckland.

Charles Lyell,
Historian of Time's Cycle

The Case of Professor Ichthyosaurus

Few scientists are so full of fun and color that their anecdotes outlive their ideas. Yet professors of geology still tell stories about the Reverend William Buckland (1784–1856) who ended his career as the prestigious Dean of Westminster, but began as England's first great academic geologist, reader at Oxford, and teacher of Charles Lyell, among others. Remember the time Buckland identified the ever-liquefying "martyr's blood" on the pavement of a continental cathedral as bat urine—by the most direct method of kneeling down and having a lick. And, oh yes, what about the day that he served crocodile meat for breakfast at the deanery, after horse's tongue the night before. Even the ever-genial Charles Darwin professed a distaste for Buckland, "who though very good humored and good-natured seemed to me a vulgar and almost coarse man. He was incited more by a craving for notoriety, which sometimes made him act like a buffoon, than by a love of science."

When Buckland was commissioned to write one of the Bridge-water Treatises "on the power, wisdom and goodness of God, as manifested in the creation," he devoted a chapter to the ichthyosaur as a primary illustration of divine benevolence. He presented all the conventional arguments for inferring God's handiwork from the anatomical perfection of this oddly fishlike reptile—"these deviations [from ordinary reptilian form] are so far from being fortuitous, or evidencing imperfection, that they present examples of perfect appointment and judicious choice . . . We cannot but recognize throughout them all, the workings of one and the same eternal principle of Wisdom and Intelligence, presiding from first to last over the total fabric of Creation" (1836, 1841 ed., 145–146). Yet Buckland could never bypass his fascination with odd illustrations of nether ends. So he devoted an even longer section to the corroboration of good design provided by the structure of an ichthyosaur's invisible intestine as inferred from the form of its coprolites, or fossil feces—reveling in the proof of God's immense care and attention to detail, as provided by the "beneficial arrangements and compensations, even in those perishable, yet important parts" (154).

Frank Buckland followed his father's footsteps in girth, conviviality, and zoophagy.[1] He was also the foremost popularizer of natural history in England, a David Attenborough for the 1850s. Frank found among his father's papers a copy of a remarkable lithograph (Figure 4.1) drawn by Sir Henry De la Beche, English to the core despite his francophonic name, and first director of the British Geological Survey. This celebrated lithograph (made so by Frank's publication as the frontispiece to his four-volume collection, *Curiosities of Natural History*) shows Professor Ichthyosaurus, surrounded by a crowd of attentive students of that ilk, and lecturing upon a peculiar fossil from ancient times—a human skull. Immediately we grasp the incongruity and the humor. We do not contemplate an ancient Jurassic ichthyosaurus, shedding its coprolites into the waters of Lyme Regis, but a *future* Professor Ichthyosaurus lecturing on the ancient stratigraphy of our present day. De la Beche's title affirms this interpretation: "Awful changes. Man found only in a fossil state. Reappearance of Ichthyosauri."

1. Not so bizarre a hobby as it might seem today. Naturalists in the heyday of Victorian expansionism hoped that the game herds of Asia and Africa might be domesticated to advance the British palate beyond beef and mutton. A systematic gustatory survey seemed both adventurous and potentially useful.

Since William Buckland had lectured so often on these beasts, and since he and De la Beche had been fast friends, Frank made the reasonable inference that this lithograph had been drawn for his father, and that the bedecked professor represented Buckland himself. Frank wrote in the preface to the first edition of his *Curiosities*:

> The frontispiece . . . is . . . a drawing made many years ago for Dr. Buckland by the late lamented Sir Henry de la Beche. . . . It was originally, drawn as a sort of quiz upon his geological lectures at Oxford, when he was treating upon Ichthyosauri, a race of extinct fish-like lizards. The subject of the drawing may be thus described—Times are supposed to be changed. Man is found only in a fossil state, in the same condition as the ichthyosauri are discovered at the present epoch; and instead of Professor Buckland giving a lecture upon the head of an ichthyosaurus, *Professor Ichthyosaurus* is delivering a lecture on the head of a fossil man.   (1874 ed., vii)

I bought Frank Buckland's volumes in 1970, during a sabbatical term in England; they enlivened many a train journey between Oxford and the British Museum in London. But I also remember a puzzle and a discovery, for I read Buckland's interpretation of his frontispiece and knew that he had erred. De la Beche's drawing had deeper and sharper meaning. Frank Buckland had, in ignorance of the true context, interpreted the drawing (quite naturally) as gentle and innocent fun directed toward his father. I inferred that the lithograph had to be a pointed, almost bitter barb of satire directed against the most curious passage in all three volumes of Charles Lyell's *Principles of Geology*—the book that most geologists regard as the founding document of their discipline's modern era. Lyell wrote about milder climates of a geological future:

> Then might those genera of animals return, of which the memorials are preserved in the ancient rocks of our continents. The huge iguanodon might reappear in the woods, and the ichthyo-

saur in the sea, while the pterodactyle might flit again through umbrageous groves of tree-ferns. (1830, 123)

Surely, De la Beche had drawn his future Professor Ichthyosaurus to mock this peculiar reverie. In support of this interpretation, De la Beche dated his lithograph in the lower-right-hand corner as 1830, year of publication for volume I of Lyell's *Principles*.

Martin Rudwick (1975) has since proven that Lyell, not Buckland, was the target of De la Beche's sketch. First, De la Beche did not draw the figure for Buckland, but distributed copies quite widely among his friends. More important, and conclusively, Rudwick discovered a series of satirical sketches and caricatures drawn by De la Beche in the back of a field notebook compiled during 1830 and 1831. Lyell, generally depicted as a vain theoretician in his barrister's wig, and contrasted with an honest field geologist in working clothes, is the butt of this series. The last sketch is a trial run for the final product. It depicts Professor Ichthyosaurus, the human skull, and one attentive student below, all placed in the same positions occupied in the later lithograph. Any lingering doubts about the identification with Lyell are dispersed by De la Beche's caption to the trial run: "Return of Ichthyosauri etc. 'Principles etc.'" The entire series, Rudwick infers, represents De la Beche's several attempts to create a caricature for his unhappiness with Lyell's thoughts and methods; Professor Ichthyosaurus finally satisfied him.[2]

While we rejoice in the solution to a small puzzle about a famous illustration in the history of geology, the identification of De la Beche's Professor Ichthyosaurus with Charles Lyell underscores an important paradox in the traditional reading of Lyell's role in the history of geology. The Lyell of textbook epitomes is, after all, the

2. Ironically, the evidence that initially persuaded me to identify Professor Ichthyosaurus with Lyell—the stated date of 1830—is incorrect. Rudwick shows that De la Beche made the sketch in 1831, then misremembered later when he transferred it to Solnhofen stone for lithography. Since De la Beche knew that he was satirizing a statement in Lyell's 1830 book, his error does become evidence of a different sort for the identification with Lyell.

hero of geology—a status won by wresting this discipline from the domination of vacuous, armchair, theologically tainted speculation, and establishing it as a modern science by hard reason based on empirical observation from the field. Consider, as we have in previous chapters, the assessment of Sir Archibald Geikie:

> With unwearied industry he marshalled in admirable order all the observations that he could collect in support of the doctrine that the present is the key to the past. With inimitable lucidity he traced the operation of existing causes, and held them up as the measure of those which have acted in bygone time . . . Not only did he refuse to allow the introduction of any process which could not be shown to be a part of the present system of nature, he would not even admit that there was any reason to suppose the degree of activity of the geological agents to have ever seriously differed from what it has been within human experience. (1905, 403)

If Geikie's Lyell accurately depicts the man himself, we face quite a conundrum of historical interpretation: why, in heaven's name, was the hero of rational empiricism engaging in reveries about returning ichthyosaurs and future pterodactyls right in the heart of the great work that established geology by eschewing fatuous speculation? Was Lyell merely providing some comic relief, following the principle of necessary variety known to every great dramatist? Do his returning ichthyosaurs function like the gravediggers contemplating Yorick's skull in *Hamlet,* or like the unctuous courtiers Ping, Pang, and Pong who make us forget for a moment that Princess Turandot will slay any suitor who can't answer her question? Or was Lyell perfectly serious—in which case, the traditional hagiography is dead wrong?

In this chapter I shall demonstrate that Lyell meant every word about future ichthyosaurs. Once again, our key to understanding the importance and seriousness of this passage lies in the metaphors of time's arrow and time's cycle. Lyell was no impartial empiricist, but a partisan thinker committed to defending time's cycle against a literal record that spoke strongly against a directionless world, particularly in its evidence for organic progress from fish to reptile, to mammal, to man.

The argument needs, and shall receive, more elaboration (see pages 137–142). For now, and to allay the discomfort of unresolved puzzles, I simply point out that future ichthyosaurs represent but one of Lyell's speculative forays for saving the steady state of life's complexity from a fossil record that spoke for progress. We may see, Lyell argues, an advance in design from fish to ichthyosaur to whale, but we view only the rising arc of a great circle that will come round again, not a linear path to progress. We are now, Lyell wrote just before his reveries on ichthyosaurs, in the winter of the "great year," or geological cycle of climates. Tougher environments demand hardier, warm-blooded creatures. But the summer of time's cycle will come round again, and "then might those genera of animals return . . ."

Charles Lyell, Self-Made in Cardboard

Lyell's Rhetoric

As De la Beche had noted in caricature, Charles Lyell was a lawyer by profession—a barrister no less, skilled in the finest points of verbal persuasion. Thus, although early sections in previous chapters on Burnet and Hutton treat cardboard histories as preached by textbooks, the Lyellian myth is a double whammy. The legends of Burnet and Hutton are later constructions, but Lyell built his own edifice with the most brilliant brief ever written by a scientist. This brief, moreover, established forever the cardboard history that fueled the emerging legends of Burnet and Hutton as well. Lyell constructed the self-serving history that has encumbered the study of earthly time ever since.

The first volume of Lyell's *Principles of Geology* (published in three volumes between 1830 and 1833) begins with five chapters on the history of geology and its lessons for establishing a proper approach to a modern study of the earth. Lyell's great treatise is not, as so often stated, a textbook summarizing all prevailing knowledge in a systematic way, but a passionate brief for a single, well-formed argument, hammered home relentlessly. All sections of the text, including the introductory history, push the same theme, while the order of sections also records the smoothly unfolding brief. The famous sentence penned by Darwin to introduce the last chapter of the *Origin of Species* would serve as well for Lyell's three installments: "this whole volume is one long argument."

Roughly characterized, Lyell holds that geological truth must be unraveled by strict adherence to a methodology that he did not name, but that soon received the cumbersome designation of "uniformitarianism" (in a review by William Whewell, written in 1832). Lyell captured the essence of uniformity in the subtitle to his treatise: "an attempt to explain the former changes of the earth's surface by reference to causes now in operation." The proposition seems simple enough. Science is the study of processes. Past processes are, in principle, unobservable; only their frozen results remain as evidence for ancient history—fossils, mountains, lavas, ripple marks. To learn about past processes, we must compare these past results with modern phenomena formed by processes that we can observe directly. In this sense, the present must be our key to the past (Figure 4.2).

If Lyellian uniformity only advocated this evident statement of method, it would be uncontroversial and not particularly enlightening. But Lyell held a complex view of uniformity that mixed this consensus about method with a radical claim about substance—the actual workings of the empirical world. Lyell argued that all past events—yes, every single one—could be explained by the action of causes now in operation. No old causes are extinct; no new ones have been introduced. Moreover, past causes have always operated—yes, always—at about the same rate and intensity as they do today. No secular increases or decreases through time. No ancient periods of pristine vigor or slow cranking up. The earth, in short, has always worked (and looked) just about as it does now. (I shall present a taxonomy of the various, and partly contradictory, mean-

Figure 4.2

Two illustrations from the first edition of *Principles of Geology* to show Lyell's working method of comparing ancient results with modern (and visible) causes that produce the same outcomes. Top: a modern volcano in the Bay of Naples observed in eruption during historical times. Bottom: Greek islands showing, by their topography, that they surround a volcanic vent. The large island is Santorin, favored candidate for Plato's Atlantis.

ings of uniformity, in the next section. For now, let me simply note that Lyell derived much mileage from their creative confusion.)

Lyell presents a double defense for his inclusive concept of uniformity. He unites the logical argument presented above with a historical justification rooted in an idiosyncratic view of Western science. In Lyell's historical tale, told in the Manichean tradition, forces of darkness are aligned to impede progress. The small flame of truth finally begins to flicker and, through the struggle of right-thinking men, eventually burns brighter to conquer superstition and perfidy. The forces of darkness are those men who see the past as different in form and cause; they make true science impossible and proceed only by vain speculation. Uniformity is the source of light, and progress in geology may be defined by its slow and steady growth in popularity. "A sketch of the progress of geology is the history of a constant and violent struggle between new opinions and ancient doctrines, sanctioned by the implicit faith of many generations, and supposed to rest on scriptural authority" (I, 30).[3]

To grasp Lyell's impact, we must admit a factor only reluctantly considered by scientists. Truth is supposed to prevail by force of logical argument and wealth of documentation, not by strength of rhetoric. Yet we will never comprehend the reasons for Lyell's triumph unless we acknowledge the role of his verbal skills. Science self-selects for poor writing. The profession has harbored several good writers, but very few great stylists. Charles Lyell was a great writer, and much of his enormous success reflects his verbal skills—not mere felicity in choice of words, but an uncanny ability to formulate and develop arguments, and to find apt analogies and metaphors for their support.

The premier example of Lyell's persuasion by rhetoric[4] is the famous chapter five on "causes which have retarded the progress of

3. Since I shall be quoting extensively from the three-volume first edition of Lyell's *Principles* (if only to convey some sense of the power of his prose), I shall adopt the convention of citing only the volume number and then the page. Volume I was published in 1830, volume II in 1832, and volume III in 1833.

4. A term that I use, by the way, in the literal, not the pejorative, sense.

geology." Here, Lyell's comprehensive argument by history proceeds through several stages:

1. Old and unfruitful views shared the common property of imagining (for such beliefs could be defended only by speculation) that the ancient earth operated under different causes working at different rates from modern processes—a "discordance" between past and present modes of change, in Lyell's phrase. "The sources of prejudice . . . are all singularly calculated to produce the same deception, and to strengthen our belief that the course of nature in the earlier ages differed widely from that now established" (I, 80).

2. Empirical observation of the earth permitted geologists to overcome these superstitions about dissimilar pasts.

> The first observers conceived that the monuments which the geologist endeavors to decipher, relate to a period when the physical constitution of the earth differed entirely from the present, and that, even after the creation of living beings, there have been causes in action distinct in kind or degree from those now forming part of the economy of nature. These views have been gradually modified, and some of them entirely abandoned in proportion as observations have been multiplied, and the signs of former mutations are skillfully interpreted . . . Some geologists [now] infer that there has never been any interruption to the same uniform order of physical events.   (I, 75)

3. The undoing of anti-uniformitarian superstition by geologists parallels the general path of enlightenment in human history.

> We must admit that the gradual progress of opinion concerning the succession of phenomena in remote eras, resembles in a singular manner that which accompanies the growing intelligence of every people . . . In an early stage of advancement, when a great number of natural appearances are unintelligible, an eclipse, an earthquake, a flood, or the approach of a comet, with many other occurrences afterwards found to belong to the regular course of events, are regarded as prodigies. The same delusion

prevails as to moral phenomena, and many of these are ascribed to the intervention of demons, ghosts, witches, and other immaterial and supernatural agents. By degrees, many of the enigmas of the moral and physical world are explained, and, instead of being due to extrinsic and irregular causes, they are found to depend on fixed and invariable laws. The philosopher at last becomes convinced of the undeviating uniformity of secondary causes. (I, 75–76)

Lyell then illustrates this equation of uniformity with righteousness not by actual examples but with metaphors based on ingenious thought experiments constructed to parallel real events by analogy. He invokes former belief in a young earth, correctly noting that uniformity cannot be supported by those committed to cramming our history into a few thousand years. Suppose, he argues, that an expedition led by a hypothetical Champollion discovered the monuments of ancient Egypt when Europeans thought that humans had first reached the Nile at the beginning of the nineteenth century. What would they make of the pyramids, obelisks, and ruined temples? These monuments "would fill them with such astonishment, that for a time they would be as men spellbound—wholly incapacitated to reason with sobriety. They might incline at first to refer the construction of such stupendous works to some superhuman powers of a primeval world" (I, 77).

But suppose the expedition then found "some vast repository of mummies" apparently indicating that humans had lived long ago to build these monuments. Honest observers (incipient uniformitarians) would then revise their fancies and admit that ordinary men had built the pyramids, but those committed to the old ways would have to invent even more outlandish theories to harmonize the mummies with their persistent conviction that no men had then inhabited Egypt. Lyell offers some suggestions: "As the banks of the Nile have been so recently colonized, the curious substances called mummies could never in reality have belonged to men. They may have been generated by some plastic virtue residing in the interior of the earth, or they may be abortions of nature produced by her incipient efforts in the work of creation" (I, 77).

We now begin to grasp Lyell's strategy. His story of Egypt recounts the great seventeenth-century debate about the nature of fossils. Many scientists then doubted that fossils could be remnants of organisms because the chronology of Moses was too short to support such plenitude. Theories of the *vis plastica* or *virtus formativa* then abounded.

As knowledge progresses, and all admit an earth of some antiquity, Lyell changes metaphors. We no longer deny human presence completely, but now try to compress history into far too short a time.

> How fatal every error as to the quantity of time must prove to the introduction of rational views concerning the state of things in former ages, may be conceived by supposing that the annals of the civil and military transactions of a great nation were perused under the impression that they occurred in a period of one hundred instead of two thousand years. Such a portion of history would immediately assume the air of a romance; the events would seem devoid of credibility and inconsistent with the present course of human affairs. A crowd of incidents would follow each other in thick succession. Armies and fleets would appear to be assembled only to be destroyed, and cities built merely to fall in ruins. There would be the most violent transitions from foreign or intestine wars to periods of profound peace, and the works effected during the years of disorder or tranquility would be alike superhuman in magnitude.   (I, 78–79)

Lyell also relied upon turn of phrase to convey his message. Consider some snippets from his most impassioned section—chapter 1 of volume III, with its restated epitome of the general doctrine, and its running head: "methods of theorizing in geology." Again, we read of a contrast between vain speculators, who view the past as different from the present, and patient empiricists, who uphold uniformity in modes, rates, and amounts of change. Note Lyell's pejorative descriptions of anti-uniformitarian reprobates (III, 2–3): "They felt themselves at liberty to indulge their imaginations, in guessing at what might be, rather than in inquiring what is"; "they employed themselves in conjecturing what might have been the course of nature at a remote period"; they preferred to "speculate on the possibilities of the past, than patiently to explore the realities of the present"; they "invented theories." "Never was there a dogma more calculated to foster indolence, and to blunt the keen edge of curiosity, than this assumption of the discordance between the former and the existing causes of change." Students were "taught to despond from the first." Geology could "never rise to the rank of an exact science"; it became "a boundless field for speculation." And finally, Lyell's most famous metaphor: "we see the ancient spirit of speculation revived, and a desire manifested to cut, rather than patiently to untie, the Gordian knot" (III, 6).

By contrast, consider the favorable phrases that describe uniformitarian heroes. They devote themselves to "inquiring what is," to "investigation of . . . the course of nature in their own times." They try "patiently to explore the realities of the present" by "candid reception of the evidence of those minute, but incessant mutations . . ." They have "hope of interpreting the enigmas"; they "undertake laborious inquiries" on the "complicated effects of the igneous and aqueous causes now in operation." Theirs is an "earnest and patient endeavor" (III, 2–3).

In illustrating three modes of Lyell's rhetoric—the invocation of history, the use of metaphor, and the contrast of adjectives—I have tried to show how he set up his preferred form of geology as the right (and righteous) side of a strict dichotomy between vain speculation and empirical truth, defined, respectively, as belief that causes worked differently on an ancient earth versus conviction that our planet has remained in a dynamic steady state throughout time.

The reality of history is so much more complex and interesting; the irony of history is that Lyell won. His version became a semi-official hagiography of geology, preached in textbooks to the present day. Professional historians know better, of course, but their message has rarely reached working geologists, who seem to crave these simple and heroic stories.

Modern Cardboard

If Lyell cast his version in cardboard, later retellings of the great dichotomy became even more simplistic. First of all, the two sides received names—catastrophism for the vanquished, uniformitarianism for the victors. Names wrap any remaining subtlety into neat packages. Secondly, the catastrophist position became more foolish and caricatured in the constant retelling. In particular, Lyell noted that the 5000-year time scale of Genesis had faded from respectability by 1800, and that his scientific colleagues could only stand accused (at most) of not allowing sufficient millions in their revised estimates (this switch in opinion prompted Lyell's transition in metaphors from Egyptian mummies to accelerated centuries). But later textbooks have usually blurred this distinction and imagined that the catastrophists of Lyell's *own day* still adhered to the Mosaic chronology. This ludicrous error then permits the final step—the enthronement of Lyell as the man who made geology a science by rejecting the explicitly miraculous (the only way to cram the complexities of history, as unraveled by 1830, into a mere 5000 years). Catastrophists, in short, became biblically motivated miracle-mongers, actively preventing the establishment of geology as a proper science. Later in this chapter, I shall show that catastrophists of Lyell's day were fine scientists who accepted both an ancient earth and the methodological meanings of uniformity. The notion that catastrophism implies biblical chronology is a logical error, for the two flanks of the central argument are not symmetrical. A 5000-year time scale does commit any adherent to global paroxysm as a mode of change, but belief in worldwide catastrophe need not imply a young earth. The earth might be billions of years old, and still concentrate its changes into paroxysmal moments.

For example, Loren Eiseley, in the best-known modern article about Lyell, confuses old-style biblical literalism, quite passé among scientists by 1830, with Lyell's true opponents among scientific catastrophists, and therefore depicts Lyell as the white knight of truth: "He entered the geological domain when it was a weird, half-lit landscape of gigantic convulsions, floods, and supernatural creations and extinctions of life. Distinguished men had lent the power of their names to these theological fantasies" (1959, 5).

The game of dichotomy requires standard-bearers for each side. In textbook cardboard, Georges Cuvier is Lyell's catastrophist enemy. He accepts the biblical chronology (or at least an earth of very short duration); he advocates the total extirpation of life (and its subsequent miraculous recreation) at each catastrophe; he works, probably consciously, for the church against science. What a vulgar misrepresentation! Cuvier, perhaps the finest intellect in nineteenth-century science, was a child of the French Enlightenment who viewed dogmatic theology as anathema in science. He was a great empiricist who believed in the literal interpretation of geological phenomena (see pages 132–137). His earth, though subject to intermittent paroxysm, was as ancient as Lyell's. He argued that many faunal changes following catastrophes represented migrations of preexisting biotas from distant areas. The real debate between Lyell and Cuvier, or of uniformity and catastrophism, was a grand scientific argument of substance—and its main subject was time's arrow and time's cycle. Yet consider some further textbook cardboard, from three best-sellers of the 1950s through the 1970s:

Gilluly, Waters, and Woodford (1959, 103) on Cuvier: "These [catastrophes], he believed, destroyed all existing life, and following each a whole new fauna was created: this doctrine, called Catastrophism, was doubtless in part inspired by the Biblical story of the Deluge." By the third edition (1969), their confidence had increased, and they substituted "unquestionably" for "doubtless in part."

Longwell and Flint (1969, 18): "One group of geologists, though admitting that the Earth had changed, supposed that all changes had occurred within the time scale of the biblical chronology. This meant that the changes had to be catastrophic."

Spencer (1965, 423): "Most students of the earth subscribed to the idea that the earth was no more than a few thousand years old, and that its history had been punctuated by one or more catastrophes during which all living beings had been wiped out and newly created beings followed."

Or Stokes (1973, 37), in the leading text of historical geology: "Cuvier believed that Noah's flood was universal and had prepared the earth for its present inhabitants. The Church was happy to have the support of such an eminent scientist, and there is no doubt that Cuvier's great reputation delayed the acceptance of the more accurate views that ultimately prevailed."

Finally, a new book on popular science by one of America's finest science writers: "Until Lyell published his book, most thinking people accepted the idea that the earth was young, and that even its most spectacular features such as mountains and valleys, islands and continents were the products of sudden, cataclysmic events, which included supernatural acts of God" (Rensberger, 1986, 236).

These texts then identify fieldwork as the fuel for change to uniformitarian enlightenment: "Hutton's ideas never caught on, though, until Lyell reintroduced them with massive documentation from his field studies" (Rensberger, 1986, 236). "The idea that catastrophism might be wrong may have been a reaction against theological dogmatism, but for most men it was an outgrowth of actual observation of nature" (Stokes, 1973, 37). "As geologic knowledge expanded, the job of rationalizing a growing list of events with a small but fixed supply of time became hopeless" (Longwell, Flint, and Sanders, 1969, 18).

But why does it matter? What harm is a bit of heroic folderol about an illusory past, especially if it makes us feel good about the progress of science? I would argue that we misrepresent history at our peril as *practicing* scientific researchers. If we equate uniformity with truth and relegate the empirical claims of catastrophism to the hush-hush unthinkable of theology, then we enshrine one narrow version of geological process as true *a priori,* and we lose the possibility of weighing reasonable alternatives. If we buy the simplistic idea that uniformity triumphed by fieldwork, then we will never understand how fact and theory interact with social context, and we will never grasp the biases in our own thinking (for we will simply designate our cherished beliefs as true by nature's dictates).

Moreover, I have a particular reason, in the context of this book, for fulminating against cardboard history. Once we recover Lyell's substantive objection to intelligible and intelligent catastrophism, we recognize that the real debate was not dogma versus fieldwork, but a conflict between rival empirics rooted in the theme of this book—a conflict of metaphor between time's cycle and time's arrow. Lyell was not the white knight of truth and fieldwork, but a purveyor of a fascinating and particular theory rooted in the steady state of time's cycle. He tried by rhetoric to equate this substantive theory with rationality and rectitude—and he largely triumphed. Thus, we cannot understand the importance of time's arrow and time's cycle in establishing our view of time and process until we break through this most enveloping of all cardboard histories. It shouldn't be difficult; cardboard is pretty flimsy stuff.

Lyell's Rhetorical Triumph: The Miscasting of Catastrophism

The Enigma of Agassiz's Marginalia

Louis Agassiz, the great Swiss scientist who made a home at Harvard and built the museum that I now inhabit, was a garden variety catastrophist. He developed the theory of continental glaciation, but advocated a global ice cover extirpating all life, with subsequent divine recreation. In the dichotomy that Lyell erected, Agassiz could only be an implacable opponent.

About ten years ago, I discovered Agassiz's copy of Lyell's *Principles* in the open stacks of our museum library. It contained some fascinating marginalia, incomprehensible if Lyell's great dichotomy accurately represents the geological struggles of his age. Agassiz penciled three comments in French on the margins of Lyell's preface to the third London edition of 1834, inscribed by the author to "Dr. Agassiz à Neuchâtel" (Agassiz and Lyell were close personal friends, whatever their professional disagreements).

The first two comments record just what we would expect from a catastrophist opponent. Agassiz first discusses the range in variation among modern causes, arguing by implication that Lyell had erred in ascribing large-scale events of the past to the accumulation of small, slow changes. Maintaining that many modern causes are substantial and abrupt, he pooh-poohs Lyell's claim that today's processes form a unitary set: "These causes are identical as the cause that produces good weather is identical with the cause that produces the tempest. But it never comes to anyone's mind to array them in the same category. There have always been different categories of causes."

The second comment extends this argument into the past and attacks a cardinal premise of uniformity: that phenomena of large scope arise as the summation of small changes. "But these changes, since they don't always have the same intensity in our own day, could not have so worked in former times. They have therefore differed at all times from considerable changes that never result from the addition of small changes."

So far, so good. But we then come to his summary statement, penciled on the blank left-hand page facing the beginning of Lyell's summary (Figure 4.3): "Les principes de Géologie de Mr. Lyell sont certainement l'ouvrage le plus important qui ait paru sur l'ensemble de cette science, depuis qu'elle mérite ce nom" (Mr. Lyell's *Principles of Geology* is certainly the most important work that has appeared on the whole of this science since it has merited this name). The statement is clearly in Agassiz's hand and represents his own comment, not the copied words of another reviewer (see Gould, 1979). How, after all Lyell's disparaging rhetoric, could a catastrophist praise him so highly? Something is very wrong. Either Agassiz was not a true catastrophist (but we can find none better), or he was trying to ingratiate himself (but not in private jottings), or he was either inconsistent or sarcastic (though he exhibited neither trait). I suggest another resolution: Lyell's dichotomy, widened by later textbook cardboard, is false outright. Geology in the 1830s was not a war between uniformitarian modernists and a catastrophist old guard with a hidden theological agenda.

Multiple Meanings of Uniformity and Lyell's Creative Confusion

When I was studying freshman geology as an undergraduate at Antioch College, our professor took us to a hill of travertine (limestone deposited by a spring), and told us that it was 15,000 years old, according to a principle called uniformitarianism. A colleague, he told us, had measured the current rate of accumulation. The principle of uniformity permitted us to assume that this rate had been constant—and a millimeter per year (or whatever) extrapolated to the bottom of the pile yielded an age of 15,000 years. If we do not accept the constancy of nature's laws in space and time, my professor added, we will be unable to apply any science beyond the immediate present.

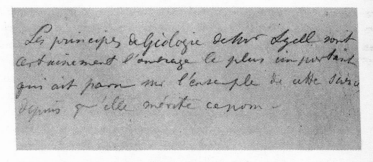

Figure 4.3
Agassiz's comment of highest praise jotted alongside his criticisms in his copy of Lyell's *Principles*.

At that stage of juvenility, I rarely challenged professorial pronouncements, but this argument seemed wrong. I could grasp the part about nature's laws, but constancy in accumulation of a travertine mound in southwest Ohio was a thing, not a principle. Why shouldn't the travertine have built twice as fast ten thousand years ago, or not at all for long stretches between periods of deposition? I turned to Lyell and to the classical sources on uniformitarianism.

Lyell, of course, never made such a crude confusion between principles and particulars, but I soon discovered that he had gathered a motley set of claims under the common umbrella of uniformity—in particular, he had made, in far more subtle form, the very same conflation of methodological principles with substantive claims that had fueled my professor's error about the travertine mound. I published my very first paper (Gould, 1965) on the multiple meanings of uniformitarianism and on the confusion that pervaded the geological literature, as debaters spoke past each other, one side invoking a meaning of uniformity to support a claim, while the other side tore it down with a different definition.

All lives contain moments of pride, and many more times better forgotten. It shall always be one of my greatest satisfactions that, as a teeny neophyte scholar alone in a little Ohio college, I noted this central confusion at the same time as a major revisionist movement to reassess Lyell's cardboard history was brewing among professional historians. Many have contributed to this revision, though I must single out the works of Hooykaas (1963), Rudwick (1972), and Porter (1976). (My own work is a belated and insignificant ripple in this tide—especially since I missed entirely the historical purpose and meaning of what Lyell had done.) Unfortunately, the message has not seeped through to practicing geologists, and textbook cardboard continues unabated.

All revisionists agree on a central point: Lyell united under the common rubric of uniformity two different kinds of claims—a set of methodological statements about proper scientific procedure, and a group of substantive beliefs about how the world really works. The methodological principles were universally acclaimed by scientists, and embraced warmly by all geologists; the substantive claims were controversial and, in some cases, accepted by few other geologists.

Lyell then pulled a fast one—perhaps the neatest trick of rhetoric, measured by subsequent success, in the entire history of science. He labeled all these different meanings as "uniformity," and argued that since all working scientists must embrace the methodological principles, the substantive claims must be true as well. Like wily Odysseus clinging to the sheep's underside, the dubious substantive meanings of uniformity sneaked into geological orthodoxy—past an undiscerning Cyclops, blinded with Lyell's rhetoric—by holding fast to the methodological principles that all scientists accepted.

We shall probably never know whether Lyell perpetrated this ruse consciously—I suspect that he did not, since his strong commitment may have engendered a personal conviction that all meanings must be true *a priori*. In any case, Lyell's rhetorical success must rank among the most important events in nineteenth-century geology—for it established an "official" history that enshrined, as the earth's own way, a restrictive view about the nature of change. If any scientist ever tries to convince you that history is irrelevant, only a repository for past errors, tell him the story of Lyell's rhetorical triumph and its role in directing more than a century of research in geology.

Using my two categories of methodological and substantive claims, Rudwick (1972) has discerned four distinct meanings of uniformity in Lyell's *Principles*.

1. *The uniformity of law.* Natural laws are constant in space and time. Philosophers have long recognized (see, in particular, J. S. Mill, 1881) that assumptions about the invariance of natural law serve as a necessary warrant for extending inductive inference into an unobservable past. (Induction, as C. S. Peirce noted, can be regarded as self-corrective in an observable present, but we can never see past processes, and no amount of current repetition can prove that present causes acted in the same way long ago—hence our need for a postulate about the invariance of nature's laws.) Or,

as James Hutton wrote with admirable directness: "If the stone, for example, which fell today were to rise again tomorrow, there would be an end of natural philosophy, our principles would fail, and we would no longer investigate the rules of nature from our observations" (1795, I, 297).

2. *The uniformity of process.* If a past phenomenon can be rendered as the result of a process now acting, do not invent an extinct or unknown cause as its explanation. This principle bears the confusing name "actualism" in reference to the meaning of its cognate (*actualisme, Aktualismus*) in most continental languages—where actual means "present," not "real" as in English. Hence, actualism is the notion that we should try to explain the past by causes now in operation. Philosopher Nelson Goodman (1967) recognized that actualism is little more than geology's own way of expressing a general rule of scientific methodology, the so-called principle of simplicity: don't invent extra, fancy, or unknown causes, however plausible in logic, if available processes suffice.

These two meanings of uniformity are statements about methodology, not testable claims about the earth. You can't go to an outcrop and observe either the constancy of nature's laws or the vanity of unknown processes. It works the other way round: in order to proceed as a scientist, you assume that nature's laws are invariant and you decide to exhaust the range of familiar causes before inventing any unknown mechanisms. Then you go to the outcrop. The first two uniformities are geology's versions of fundamental principles—induction and simplicity—embraced by all practicing scientists both today and in Lyell's time.

But Lyell's other uniformities are radically different in status. They are testable theories about the earth—proposals that may be judged true or false on empirical grounds.

3. *Uniformity of rate, or gradualism.* The pace of change is usually slow, steady, and gradual. Phenomena of large scale, from mountain ranges to Grand Canyons, are built by the accumulation, step by countless step, of insensible changes added up through vast times to great effect (Figure 4.4). Major events do, of course, occur—

especially floods, earthquakes, and eruptions. But these catastrophes are strictly local; they neither occurred in the past, nor shall happen in the future, at any greater frequency or extent than they display at present. In particular, the whole earth is never convulsed at once, as some theorists held. Speaking of floods, for example, Lyell writes:

> They may be introduced into geological speculations respecting the past, provided we do not imagine them to have been more

The dotted line represents the sea-level.

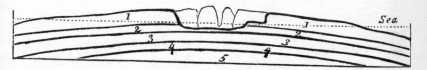

Figure 4.4

A classic example of Lyell's gradualism—the "denudation of the Weald." The geological structure of this eroded basin (top) is a great dome, bowed up after deposition of the Chalk in Cretaceous times (number 1 in both figures). The bottom figure represents the dome soon after its formation, with breakage and erosion just beginning in chalk deposits at the dome's summit. The top figure shows that, by gradual erosion of the Chalk and underlying layers, the modern basin has formed, with a substantial separation (by denudation) of the North and South Downs (the upstanding sides of the original dome, labeled as number 1). Lyell's premier example of gradualism led Darwin into one of his most famous errors. In the first edition of the *Origin of Species*, he used the denudation of the Weald to illustrate the insensible slowness of gradual change in geology. He estimated the time required for this denudation as 300 million years (though the actual time since deposition of the Chalk is only 60 million years or so). In later editions, Darwin dropped this calculation after receiving severe criticism. (Illustrations from the first edition of Lyell's *Principles*.)

frequent or general than we expect them to be in time to come. (I, 89)

Lyell defends the third uniformity with his characteristic ploys of rhetoric. He marshals, for example, his whiggish view of historical progress to assert that lingering notions of worldwide paroxysm are vestiges of a savage past when men huddled in fear before the thunderbolt.

> The superstitions of a savage tribe are transmitted through all the progressive stages of society, till they exert a powerful influence on the mind of the philosopher. He may find, in the monuments of former changes on the earth's surface, an apparent confirmation of tenets handed down through successive generations, from the rude hunter, whose terrified imagination drew a false picture of those awful visitations of floods and earthquakes, whereby the whole earth as known to him was simultaneously devastated. (I, 9)

The replacement of catastrophe by accumulated slow change is the very essence of progress:

> The mind was slowly and insensibly withdrawn from imaginary pictures of catastrophes and chaotic confusion, such as haunted the imagination of the early cosmogonists. Numerous proofs were discovered of the tranquil deposition of sedimentary matter and the slow development of organic life. (I, 72)

(Note how the progress of mind mirrors nature's own way—"slowly and insensibly.")

Catastrophism, by its very nature, is anti-empirical:

> Instead of confessing the extent of their ignorance, and striving to remove it by bringing to light new facts, they would be engaged in the indolent employment of framing imaginary theories concerning catastrophes and mighty revolutions in the system of the universe. (I, 84)

Thus, we may reject, as unintelligible in principle, the idea of worldwide, or even substantial regional catastrophe, for it would be

contrary to analogy to suppose, that Nature had been at any former epoch parsimonious of time and prodigal of violence— to imagine that one district was not at rest while another was convulsed—that the disturbing forces were not kept under subjection, so as never to carry simultaneous havoc and desolation over the whole earth, or even one great region.   (I, 88–89)

These five quotations illustrate Lyell's debating style. He establishes certain meanings of uniformity—law and process—as necessary postulates of scientific method, and then tries to win similar status for controversial ideas about the earth's empirical behavior by describing them in the same language of logical necessity.

4. *Uniformity of state, or nonprogressionism.* Change is not only stately and evenly distributed throughout space and time; the history of our earth also follows no vector of progress in any inexorable direction. Our planet always looked and behaved just about as it does now.[5] Change is continuous, but leads nowhere. The earth is in balance or dynamic steady state; therefore, we can use its current *order* (not only its laws and rates of change) to infer its past. Land and sea, for example, change positions in an endless dance, but always maintain about the same proportions. Floods, volcanoes, and earthquakes have wrought devastation at about the same frequency and extent throughout time. The earth had no early period of more vigorous convulsion.

Lyell also extended the uniformity of state to life. Species are real entities with points of origin in space and time, definite durations and moments of extinction. Their beginnings and ends are not

5. Like Hutton, Lyell eschewed speculation about ultimate beginnings and ends. These moments must, of course, depart from uniformity of state, but science cannot comprehend them. The uniformities apply to the vast panorama of time as recorded in the geological record.

concentrated in episodes of mass death or radiation, but are distributed evenly through space and time—another dynamic steady state as introductions are balanced by removals. Moreover, the timing of introductions displays no progress in organization or complexity, no advance through the chain of being. Lyell argued (see pages 137–142) that appearances of progress in stratigraphic origins (from fish to reptile to mammal to human in the history of vertebrates, for example) were illusory. He believed that mammals had lived during earliest Paleozoic times and that future exploration would recover their fossil remains from these ancient rocks.

Lyell's extension of the fourth uniformity to life strikes many people as intensely puzzling. They can appreciate the force behind a claim for uniformity of state in the physical world, but surely we know that life must change in a progressive manner. Yet, for Lyell, the link of life to the physical world was direct and necessary. He held that species arose (whether by God's hand or by some unknown secondary cause) in perfect adaptation to physical surroundings. Any progressive change in life's state could only mirror a corresponding alteration in the physical environment. Since uniformity of state pervades climate and geography, life must also participate in the nondirectional dance.

Lyell defended uniformity of state with the same devices of rhetoric that he had applied to gradualism—he conflated this testable and controversial theory about the nature of things with methodological canons that all scientists accept, thereby attempting to secure an *a priori* status for time's cycle as a necessary component of rationality itself.

Consider two examples. Lyell often argued against directional theories not by citing contrary evidence, but by holding that their claims for a different earth in the past could not be rendered accessible to inquiry or even made intelligible. He refutes the old Neptunian theory of original deposition from a universal ocean not with stratigraphic evidence, but in principle because it advocates an ancient earth different in state from our own:

If, at certain periods of the past, rocks and peculiar mineral composition had been precipitated simultaneously upon the floor of an "universal ocean," so as to invest the whole earth in a succession of concentric coats, the determination of relative dates in geology might have been a matter of the greatest simplicity. To explain, indeed, the phenomenon would have been difficult, or rather impossible, as such appearances would have implied a former state of the globe, without any analogy to that now prevailing. (III, 37–38)

In the most striking of all Lyellian statements, he sketches a series of tactics to preserve uniformity of state in the face of almost any conceivable evidence for directional change in the earth's history:

When we are unable to explain the monuments of past changes, it is always more probable that the difficulty arises from our ignorance of all the existing agents, or all their possible effects in an indefinite lapse of time, than that some cause was formerly in operation which has ceased to act; and if in any part of the globe the energy of a cause appears to have decreased, it is always probable, that the diminution of intensity in its action is merely local, and that its force is unimpaired, when the whole globe is considered. But should we ever establish by unequivocal proofs, that certain agents have, at particular periods of past time, been more potent instruments of change over the entire surface of the earth than they now are, it will be more consistent with philosophical caution to presume, that after an interval of quiescence they will recover their pristine vigor, than to regard them as worn out. (I, 164–165)

I have quoted this passage for years, but it never ceases to astound me. It begins with an acceptable statement of the second uniformity, then slides into substantive claims about uniformity of state. Finally, Lyell actually holds that we should reject directional change even if we have "unequivocal proofs . . . over the entire surface of the earth," because we have a right to anticipate that any apparently depleted cause will, in future, resume its former intensity.

Will the Real Catastrophism Please Stand Up:
The Solution to Agassiz's Paradox

This exegesis can resolve the paradox of Agassiz's private reactions to Lyell's brief. Agassiz did not view the professional world of geology in 1830 as a battleground between scientific empiricists and theological apologists. As a fellow scientist, Agassiz accepted the methodological uniformities of law and process. As a catastrophist who viewed the history of vertebrates as a tale of progress, he rejected the substantive uniformities of rate and state. On balance, praise for Lyell's forceful and beautifully crafted defense of proper method far exceeded unhappiness with their longstanding disagreement about the earth's behavior—for scientists then and now have recognized that their profession is defined by its distinctive modes of inquiry, not by its changing perceptions of empirical truth.

As we have our von Danikens, our "scientific" creationists, and our faith healers, science in Lyell's day also felt besieged by a periphery of charlatans and reactionaries, who often commanded much public support. Agassiz therefore welcomed most heartily Lyell's elegant defense of scientific method—this "most important work . . . since [geology] has merited this name." Once we abandon the cardboard version of Lyell's scientific light versus Agassiz's theological darkness, and grasp their substantial common ground by their own perceptions, Agassiz's general praise following his particular criticisms provokes no surprise. Lyell's dichotomy, exaggerated by subsequent textbook cardboard, presents a taxonomy that would not have been accepted, or even recognized, by most of his contemporaries. In Agassiz's classification, any primary division between supporters and detractors of fruitful science would have placed him and Lyell on the same side.

As evidence of support by catastrophists for the methodological uniformities of law and process, consider two primary "villains" of the cardboard version—men usually depicted as enemies of modern science. I have already demonstrated in Chapter 2 that Burnet's strict adherence to uniformity of law not only defined his approach to the earth but also, and ironically, directly inspired the fanciful conjectures that have set his poor reputation in modern textbooks. Burnet accepted the Bible as literal truth. If he had been willing to admit miracles as agents of earthly change, his explanations would have been no more curious than those of Genesis. But Burnet was committed, by uniformity of law, to rendering even the most elaborate of biblical tales by the workings of Newtonian physics—and such an effort required some mighty tall and fancy conjectures about sources of water and formation of topography. Still, Burnet refused, on methodological grounds, to permit divine creation of floodwaters, and even used the metaphor of Lyell's most famous passage (see page 111) to defend the uniformity of law: "They say in short, that God Almighty created waters on purpose to make the deluge . . . And this, in a few words, is the whole account of the business. This is to cut the knot when we cannot loose it" (Burnet, 33).

The French geologist Alcide d'Orbigny usually wins a textbook nod as most outlandish of catastrophists among Lyell's contemporaries. He identified some twenty-eight episodes of global paroxysm, marked by volcanoes, tidal waves, and effusion of poison gases, and leading to the annihilation of life. Yet d'Orbigny embraced the actualist principle of uniformity of process. He recommended that inquiry always begin with modern processes: "Natural causes now in action have always existed . . . To have a satisfactory explanation of all past phenomena, the study of present phenomena is indispensable" (1849–1852, 71). And he followed Agassiz in praising Lyell for this proper emphasis (though he reserved greatest honors for a fellow countryman): "The happy thought that we should explain the earth's strata by causes now acting belongs entirely to Mr. Constant Prevost who first established it in his geological system. Science owes much to Mr. Lyell for the development of this system, supported by copious research as wise as it is ingenious" (1849–1852, 71).

D'Orbigny and Agassiz had no beef with Lyell about method—they all agreed that inquiry must begin with present processes and that modern causes must be entirely exhausted before any extinct or exotic process be considered. They differed in their judgment about the world's response to this common method. Lyell believed that modern causes would suffice to explain everything about the past. Agassiz and d'Orbigny viewed actualism as a method of subtraction for identifying the unchanging substrate of modern causes, thereby highlighting the phenomena that required special explanation by processes outside the current range.

Nonetheless, d'Orbigny also admitted that modern causes would greatly aid the understanding of past catastrophes—for paroxysms may be caused by modern forces greatly magnified in degree. D'Orbigny argued, for example, that any change in topography during an earthquake "is, for us, on a small scale, and with effect much less marked, the same phenomenon as one of the great and general perturbations to which we attribute the end of each geological epoch" (1849–1852, II, 833–834).

All agreed, therefore, that the most valuable of all possible tools for interpreting the past would be a proper catalogue of the variety, range, rates, and extent of modern causes. Lyell had provided the finest compendium ever assembled. I believe that this elaborately detailed catalogue, above all else, won for Lyell the bounteous praise that catastrophists like Agassiz gratefully accorded.

Thus, the real debate between Lyell and the catastrophists was a complex argument of substance among men who agreed about methods of inquiry. Lyell's substantive uniformities of rate and state commingle to form a powerful vision of a dynamic earth, constantly in motion but never changing in general appearance or complexity—a stately earth playing a modest portion of all its acts all the time, not concentrating single modes of change into global episodes, and not alternating worldwide periods of tumult and quiescence, uplift and erosion.

Lyell's attitude to James Hutton illustrates the mixture of his substantive uniformities. Lyell praised Hutton for his nondirectional world machine with endlessly cycling phases of uplift, erosion, deposition, consolidation, and uplift—an earthly mechanics beautifully consonant with the fourth uniformity of state. But Lyell criticized Hutton for viewing the stages of his cycle as a global succession and, especially, for the catastrophic character of his periods of uplift—a violation in the third uniformity of rate. He cited as "one of the principal defects" of Hutton's theory "the assumed want of synchronism in the action of the great antagonist powers—the introduction, first, of periods when continents gradually wasted away, and then of others when new lands were elevated by violent convulsions" (II, 196). Lyell insisted upon stately unfolding going nowhere.

No logical connection unites the two substantive uniformities of rate and state. One could, like Hutton, accept nondirection while advocating catastrophic periods of uplift. But Lyell's vision joined them neatly and tightly. I shall subsequently refer to Lyell's junction of rate and state, the essence of his world view, as "time's stately cycle."

Similarly, as "catastrophism" confutes by its name only the third uniformity (of rate), pretenders to this title might espouse paroxysm without direction. But all prominent catastrophists of Lyell's day also linked the uniformities of rate and state—by denying them both. From Cuvier, d'Orbigny, and Elie de Beaumont in France, to Agassiz in Switzerland (and then America), to Buckland and Sedgwick in England, they agreed that occasional paroxysm had been the predominant mode of substantial change on an ancient earth. These catastrophes occurred as direct consequences of the primary and inherent directionality that had also provoked the progressive increase in life's complexity—cooling of the globe. So tight was this link between catastrophe and direction that Martin Rudwick and other historians prefer to designate this theory as "the directionalist synthesis."

Since the catastrophists of Lyell's day were fine scientists, not the vestigial miracle-mongers that Lyell described in his rhetoric, their characteristic linkage of catastrophe and direction rested upon respected theories of physics and cosmology. In simplified essence, the earth had formed hot (in a molten or gaseous state), as maintained by the nebular hypothesis of Kant and Laplace, then the leading theory for the origin of our solar system. As the physics of large bodies dictates, the earth had cooled steadily through time. As the earth cools, it contracts. The outer crust solidifies, but the molten interior continues to shrink and "pull away" from the rigid surface. This contraction creates an instability that becomes more and more severe until the rigid crust cracks and collapses upon the shrunken core. The earth's intermittent paroxysms are these geological moments of violent readjustment—and they explain a host of empirical phenomena, including the linearity of mountain chains as cracks of shrinkage or breakage. Since life adapts to environment, the harsher worlds of our cooling earth have engendered more complex creatures better able to cope.

How did Lyell respond to this powerful theory of the earth and life? In part, he reacted as the stated ideals of science profess—by defending his own vision with evidence and theoretical arguments (see the next section of this chapter). But he also counterattacked with the same successful rhetoric that secured the cardboard history of geology—he again conflated phenomena and procedures, arguing that the substantive claims of catastrophism are unintelligible in principle because all scientists accept the methodological uniformities of law and process.

Lyell begins his attack on Elie de Beaumont's physics of catastrophism in the historical introduction to his chapter "on the causes of vicissitudes in climate" (I, ch. 7). Lyell identifies his usual bugbear of different causes on an ancient earth as the wrong way to understand geological changes in climate. He designates as villain "the cosmogonist" who "has availed himself of this, as of every obscure problem in geology, to confirm his views concerning a period when the laws of the animate and inanimate world were wholly distinct from those now established" (I, 104). As illustrations, Lyell chooses the most outlandish of old and discredited ideas, especially Burnet's change in axial tilt following Noah's flood. He then makes a quick transition from these abandoned fancies to the respectable notion of directional cooling from an original molten state, but manages to place this physical basis of contemporary catastrophism into the same pot with discredited cometary collisions, thereby branding the best physics of his day as vain speculation. He then rejects directional cooling *a priori*, on methodological grounds:

> When the advancement of astronomical science had exploded this theory [Burnet's axial changes], it was assumed that the earth at its creation was in a state of fluidity, and red hot, and that ever since that era it had been cooling down, contracting its dimensions, and acquiring a solid crust—an hypothesis equally arbitrary, but more calculated for lasting popularity, because, by referring the mind directly to the beginning of things, it requires no support from observations, nor from any ulterior hypothesis. They who are satisfied with this solution are relieved from all necessity of inquiry into the present laws which regulate the diffusion of heat over the surface, for however well these may be ascertained, they cannot possibly afford a full and exact elucidation of the internal changes of an embryo world. (I, 104–105)

Lyell then presents his own recommendation:

> But if, instead of vague conjectures as to what might have been the state of the planet at the era of its creation, we fix our thoughts steadily on the connection at present between climate and the distribution of land and sea . . . we may perhaps approximate to a true theory. If doubt still remain, it should be ascribed to our ignorance of the laws of Nature, not to revolutions in her economy;—it should stimulate us to farther research, not tempt us to indulge our fancies in framing imaginary systems for the government of infant worlds. (I, 105)

Later (III, ch. 24), Lyell attacks Elie de Beaumont because his "successive revolutions . . . cannot be referred to ordinary volcanic forces, but may depend on the secular refrigeration of the heated interior of our planet" (III, 338–339). Lyell reasserts his preference for shifting foci of internal volcanic heat, constant in strength through time, but moving from place to place on the earth's surface—for this idea presumes "the reiterated recurrence of minor convulsions" rather than unacceptable "paroxysmal violence" (III, 339). Lyell attacks Elie de Beaumont not with facts that support his uniformity of state, but with a claim that directionalism is unscientific ("mysterious in the extreme"):

> The speculation of M. de Beaumont concerning the "secular refrigeration" of the internal nucleus of the globe, considered as a cause of the instantaneous rise of mountain-chains, appears to us mysterious in the extreme, and not founded upon any induction from facts; whereas the intermittent action of subterranean volcanic heat is a known cause capable of giving rise to the elevation and subsidence of the earth's crust without interruption of the *general* repose of the habitable surface. (I, 339)

But what is inherently preferable about causes that preserve the general repose of the surface?

Lyell and the catastrophists were locked in a fascinating debate of substance about the way of our world, not a wrangle about methodological aspects of uniformity. Their struggle pitted a directional view of history as a vector leading toward cooler climates and more complex life, and fueled by occasional catastrophes, against Lyell's vision of a world in constant motion, but always the same in substance and state, changing bit by bit in a stately dance toward nowhere. This real debate, so lost at our peril in the success of Lyell's rhetoric, was the grandest battle ever fought between the visions of time's arrow and time's cycle.

Lyell's Defense of Time's Cycle

Lyell's Distinctive Method of Probing behind Appearances

Lyell's work may be awash in rhetoric but it is, as Agassiz fairly noted, an intellectual *tour de force* filled with meatier arguments of great interest.

Lyell and his catastrophist opponents differed not only in their interpretation of the geological record, but also in their basic approach to field evidence. In the light of Lyell's rhetorical brand, marking his opponents as anti-empiricists devoted to armchair speculation, their differences in approach present one of the great ironies in the history of science.

Read literally, then and now, the geological record is primarily a tale of abrupt transitions, at least in local areas. If sediments indicate that environments are changing from terrestrial to marine, we do not usually find an insensibly graded series of strata, indicating by grain size and faunal content that lakes and streams have given way to oceans of increasing depth. In most cases, fully marine strata lie directly atop terrestrial beds, with no signs of smooth transition. The world of dinosaurs does not yield gradually to the realm of mammals; instead, dinosaurs disappear from the record in apparent concert with about half the species of marine organisms in one of the five major mass extinctions of life's history. Faunal transitions, read literally, are almost always abrupt, both from species to species[6] and from biota to biota.

The characteristic method of catastrophism, promulgated particularly by Cuvier, was empirical literalism—an approach diametrically opposed to Lyell's unfair characterization of these scientists as speculators opposed to field evidence. The catastrophists tended to accept what they saw as reality: abrupt transitions of sediments and fossils indicated rapid change of climates and faunas. The defense of catastrophism was rooted in the most direct (or minimally "interpretive") reading of geological evidence.

Lyell did not deny this apparent evidence of abruptness—that is, he did not defend the uniformity of rate by citing different direct evidence for gradual transitions. He couldn't, since the literal record speaks too loudly for discontinuity. Instead, he supported uniform-

---

6. Niles Eldredge and I developed the theory of punctuated equilibrium to explain these transitions between species as an accurate reflection of the workings of evolution, not as artifacts of an imperfect fossil record.

ity of rate with a brilliant argument for "probing behind" literal appearances, and trying to find the signal of true gradualism in a record so riddled with systematic imperfections that insensible transitions become degraded to bits and pieces of apparent abruptness.

I do not gloat in this analysis to "show up" Lyell as less empirical or field-oriented than his catastrophist opponents. I find no particular virtue in empirical literalism and generally support Lyell's approach for balancing fact and theory in a complex and imperfect world. I just find it deliciously ironic that cardboard history touts Lyell's victory as the triumph of fieldwork, while catastrophists were the true champions of a geological record read as directly seen. Lyell, by contrast, urged that theory—the substantive uniformities of rate and state—be imposed upon the literal record to interpolate within it what theory expected but imperfect data did not provide.

Early in his first volume, Lyell admits the literal appearance of catastrophe as predominant in geology:

> The marks of former convulsions on every part of the surface of our planet are obvious and striking . . . If these appearances are once recognized, it seems natural that the mind should come to the conclusion, not only of mighty changes in past ages, but of alternate periods of repose and disorder—of repose when the fossil animals lived, grew, and multiplied—of disorder, when the strata wherein they were buried became transferred from the sea to the interior of continents, and entered into high mountain chains.   (I, 7)

To uphold the third uniformity (of rate) in the face of this admission, Lyell uses two arguments, both based on probes "behind appearance." First, he argues that local records cannot be extrapolated to wider regions, or to the whole globe. Abrupt transition in one section, for example, may be reconciled with a world in balance if we find opposite changes in other places at the same time. "There can be no doubt, that periods of disturbance and repose have followed each other in succession in every region of the globe, but it may be equally true, that the energy of the subterranean movements has been always uniform as regards the whole earth" (I, 64).

Second, using his metaphor of the book (later adopted by Darwin for defending gradualism in fossil sequences), Lyell argued that slow and continuous change will degrade to apparent abruptness as fewer and fewer stages are preserved—as if, of the original book, an imperfect record preserved but few pages, of the pages few lines, of the lines few words, and of the words few letters. In Darwin's words:

> For my part, following out Lyell's metaphor, I look at the natural geological record, as a history of the world imperfectly kept, and written in a changing dialect; of this history we possess the last volume alone, relating only to two or three countries. Of this volume, only here and there a short chapter has been preserved; and of each page, only here and there a few lines.   (1859, 310–311)

Lyell expands this central theme of imperfection in a double metaphor (III, ch. 3). Compare the continuous origin and extinction of species with birth and death in human populations. Let preservation in geological strata correspond to the records of census takers. The appearance of true gradualism or illusory catastrophe then depends upon the density of preserved information. If a nation contains sixty provinces and a full census be taken every year in each, then the preserved record will match the actual character of slow and continuous change. But suppose that an impoverished or distracted government can employ only one team of census takers, who can, at best, visit only one province a year. Then the records for each province, spaced sixty years apart, will display an almost complete turnover of population from one recorded instance to the next—the illusory catastrophe that appears when continuity is sampled too sparsely. Of course, geological records are even more scanty and erratic. The geological "census takers" do not sample in strict rotation; some areas may remain "unvisited" for great stretches of time, while records duly made and entered may be destroyed by

subsequent erosion. Lyell concludes that the literal record of catastrophic faunal turnover really represents a continuous change of
life filtered through ordinary laws of sporadic sedimentation: "If
this train of reasoning be admitted, the frequent distinctness of the
fossil remains, in formations immediately in contact, would be a
necessary consequence of the existing laws of sedimentary deposition, accompanied by the gradual birth and death of species" (III,
32–33).

Lyell then switches metaphors to illustrate the important corol-
lary that signs of disturbance in an illusory transition need bear no
relation to the actual causes of change. Suppose that a modern
eruption of Mt. Vesuvius buried an Italian city atop Herculaneum.
The abrupt change in language and architecture, as seen in the
archaeological record, would not only be illusory, but also quite
unrelated to the catastrophe of volcanic eruption.

As Lyell defended gradualism by probing behind the literal appearance of catastrophe, he supported the second substantive uniformity (of state) with a similar admission and resolution. He also
granted that several vectors of directional change ran through the
geological record read literally—older rocks tend to be denser,
harder, and more altered by heat and pressure; climates (at least in
the northern hemisphere) have become harsher, as indicated by
sediments and the fossils they contain; life itself (at least for vertebrates) has become more complex. Lyell argued that each apparent
vector is an illusion produced by directional biases of preservation
acting upon a uniform world in steady state.

Lyell again relied upon metaphor to express these (then) unfamiliar and crucial arguments. Suppose that a collector of insects
shipped specimens from a tropical land to England, with a minimum transport of two months, and suppose that these organisms
lived little longer than two months (and did not breed in captivity).
Englishmen would then see only aged adults. Likewise, old rocks
are often contorted and metamorphosed, younger rocks evenly lay-
ered and less dense. Many geologists had viewed this directional
change in strata as a sign of decreasing intensity in geological forces,

perhaps the signature of a cooling earth. But Lyell argued, by his entomological metaphor, that forces of uplift and consolidation might be unvarying through time, as the uniformity of state required. The older the rock, the longer it might be subject to constant forces of alteration, and the more it might become baked and contorted. Only old rocks are so altered, just as all insects reaching England are adult—but as beetle life cycles flow from larva to adult in their native land, so too are rocks continually made in the bowels of the earth, but only receive the imprint of contortion and metamorphosis as they move toward the surface through time: "If the disturbing power of the subterranean causes be exerted with uniform intensity in each succeeding period, the quantity of convulsion undergone by different groups of strata will generally be great in proportion to their antiquity" (III, 335). Direction, in other words, is an illusion, as older rocks receive more "attention" from the constant forces of time's cycle.

The Worst Case as Crucial Test: Lyell Probes Behind Appearance to Deny Progression in Life's History

Most grand visions have crucial tests or tragic flaws. The paleontological record played this dual role as goad and bugbear throughout Lyell's career as he attempted to validate his vision of time's stately cycle. The problem is simply stated: no other aspect of geology seems so clearly progressive in our usual, vernacular sense—especially given our inordinate interest in ourselves, our smug convictions about human superiority, and the restriction of human fossils to the last microsecond of geological time.

The invertebrate record might easily be read in the light of time's cycle—since most anatomical designs first appear at roughly the same time in the oldest fossiliferous strata (as known in Lyell's time). But how could appearances of progress—at least in the parochial sense of increasing taxonomic proximity to *Homo sapiens*—be denied as reality in the record of vertebrates? Fish came first, then reptiles, then mammals, and finally human artifacts at the very

top of the stratigraphic pile. Paleontologists had searched assidu-
ously for vertebrate remains. Could this literal appearance also be
ascribed to incompleteness of the record? Lyell quotes Sir Humphrey Davy on literal appearance as confuting uniformities of both
rate and state: "There seems, as it were, a gradual approach to the
present system of things, and a succession of destructions and
creations preparatory to the existence of man" (I, 145).

Lyell responds to this greatest challenge in chapter 9 of volume
I: "theory of the progressive development of organic life considered—evidences in its support wholly inconclusive." He divides the
attack on uniformity into two separate questions requiring different
responses: "First, that in the successive groups of strata, from the
oldest to the most recent, there is a progressive development of
organic life, from the simplest to the most complicated forms; —
secondly, that man is of comparatively recent origin" (I, 145). The
first claim, he argues, "has no foundation in fact"; the second,
though "indisputable," is not "inconsistent with the assumption,
that the system of the natural world has been uniform . . . from the
era when the oldest rocks hitherto discovered were formed" (I,
145).

Lyell uses two kinds of arguments to refute the first claim, that
vertebrates march up life's ladder in stratigraphic order. These arguments may not stand formally in contradiction; but they certainly
illustrate Lyell's willingness to exploit both sides of a potential
weakness.

First argument. Advanced vertebrates were present in the earliest
strata as well, but we haven't found their remains yet. Lyell here
invokes his most characteristic argument—the appearance of progress is caused by directional biases in preservation, not by progressive trends in actual history. First of all, apparent progress is not
so marked or pervasive. Complex fish appear in the earliest strata,
reptiles soon after, and still in old rocks (now called Paleozoic).
The evidence for progress is entirely negative—the absence only of
birds and mammals in Paleozoic rocks. Birds so rarely fossilize that
biases of better preservation in more recent rocks might restrict

their remains to later strata even if they had actually lived at modern abundances in Paleozoic times.

Lyell could not evoke the same argument for mammals—since their dense and massive bones fossilize more easily. He therefore invoked two biases of discovery to argue that Paleozoic mammals abounded, but have not yet been found as fossils. Our explorations have been largely restricted to Europe and North America, a small segment of the globe. This region was the center of an ocean during the Paleozoic, far from any continent that might yield a floating carcass to full fathom five. After all, we might dredge an equally large area of the central Pacific today, and find no signs of mammalian life:

> The casualties must be rare indeed whereby land quadrupeds are swept by rivers and torrents into the sea, and still rarer must be the contingency of such a floating body not being devoured by sharks . . . But if the carcass should escape and should happen to sink where sediment was in the act of accumulating, and if the numerous causes of subsequent disintegration should not efface all traces of the body included for countless ages in solid rock, is it not contrary to all calculation of chances that we should hit upon the exact spot—that mere point in the bed of the ancient ocean, where the precious relic was entombed?   (I, 149)

But Lyell's trump card was an empirical discovery, not a verbal argument. Thirty years before, the fossil record of mammals had provided even better signs of apparent progress—for their remains had been entirely confined to the latest, or Tertiary, rocks. None had been found throughout the entire middle realm, now called Mesozoic and popularly known as the age of dinosaurs. But, by 1830, a few small mammals had been discovered in the midst of Mesozoic strata. If the Mesozoic had fallen to assiduous exploration, could the older Paleozoic rocks be far behind?

Second argument. Perhaps advanced vertebrates really did not live during the early ages of fishes and primitive reptiles, but their absence is a contingent and reversible consequence of climatic change, not the mark of an inexorable vector of progress.

Following the old warrior's advice that one must be prepared for all contingencies, Lyell girds himself to defend time's cycle even if Paleozoic mammals are never found, and the vector of apparent progress is confirmed. If we accept Lyell's dubious premise that all species arise in perfect adaptation to prevailing environments, then the vector of progress might bear two interpretations, one fatal to time's cycle, the other consistent. Progress in vertebrate life might mean, as the catastrophists asserted, that our planet had cooled continuously, and that more complex life developed to weather the harsh decline from easy tropicality. But Lyell rejects this interpretation with his usual rhetorical flourish: "In our ignorance of the sources and nature of volcanic fire, it seems more consistent with philosophical caution, to assume that there is no instability in this part of the terrestrial system." (But why does caution urge nondirection, if vectors of cooling are consistent with the best physics and cosmology of Lyell's day?)

Or the absence of Paleozoic mammals might signify that cooling in the northern hemisphere since Paleozoic times recorded a contingent and reversible shift to more land and less sea (see pages 144–145 for details of Lyell's argument on ties of climate to relative positions and amounts of land and sea):

> We have already shown that when the climate was hottest, the northern hemisphere was for the most part occupied by the ocean, and it remains for us to point out, that the refrigeration did not become considerable, until a very large portion of that ocean was converted into land, nor even until it was in some parts replaced by high mountain chains.   (I, 134)

Trends in climate caused by the shifting dance of land and sea (rather than by inexorably cooling interiors of planets) are temporary and reversible. The uniformity of physical state suggests that any regional trend to greater continentality (and consequently increasing cold) will eventually reverse itself, since land and sea are continually changing positions, but always maintaining their relative proportions on a global scale. The northern hemisphere is now in "the winter of the 'great year,' or geological cycle"; but we may expect the future to bring "conditions requisite for producing the maximum of heat, or the summer of the same year" (I, 116).

Life, to say it once more, follows climates. If the passage from summer to winter of the great year has brought progress to vertebrate life in the northern hemisphere, the return of subsequent summer must engender a most curious result. We come then to that most stunning passage of the entire *Principles*—the line that marks Lyell as a theorist dedicated to consistency, not always to empirical restraint (as legend holds); the conjecture so *outré* (even to Lyell's contemporaries) that De la Beche captured it in caricature, while Frank Buckland, unable to grasp such a curious context, interpreted it as a jest about his father rather than a mordant dig at Lyell; the subject of the frontispiece and first section of this chapter. And so, once more with feeling: "Then might those genera of animals return, of which the memorials are preserved in the ancient rocks of our continents. The huge iguanodon might reappear in the woods, and the ichthyosaur in the sea, while the pterodactyle might flit again through umbrageous groves of tree-ferns" (I, 123).

But as zealously as Lyell probed behind appearance to impose uniformity of state upon the apparent record of vertebrate progress, he could not (or dared not) extend this argument to our own species. Humans are special; humans are different. The intellectual world is littered with systems that pushed consistency to the ends of the earth and the bounds of rationality, but then stepped aside and made an exception for human uniqueness. Lyell followed this tradition and placed a picket fence around *Homo sapiens*.

Lyell does note, quite fairly, that we often make too much of ourselves and that our physical bodies are poor and flawed indeed, displaying no mark of progress in our late appearance: "If the organization of man were such as would confer a decided pre-eminence upon him, even if he were deprived of his reasoning powers . . . he might then be supposed to be a link in a progressive chain" (I, 155).

Even our strength of reason cannot stay nature's power:

> We force the ox and the horse to labor for our advantage, and we deprive the bee of his [*sic*] store; but, on the other hand, we raise the rich harvest with the sweat of our brow, and behold it devoured by myriads of insects, and we are often as incapable of arresting their depredations as of staying the shock of an earthquake, or the course of a stream of burning lava.   (I, 162)

Lyell admits, however, that this argument can extend only so far. The late appearance of our bodies does not violate uniformity, but humans do not challenge time's cycle as mere naked apes: "The superiority of man depends not on those faculties and attributes which he shares in common with the inferior animals, but on his reason by which he is distinguished from them" (I, 155).

Although human reason is a violation of time's cycle, it is too grand, too different, too godlike for inclusion in an argument about physical and organic history. One might almost say that God made human reason at the end of time so that something conscious might delight in the grand uniformity of time's stately cycle: "No one of the fixed and constant laws of the animate or inanimate world was subverted by human agency . . . the modifications produced were on the occurrence of new and extraordinary circumstances, and those not of a physical, but a moral nature" (I, 164).

Charles Lyell was struggling, not joyfully triumphing, with his most difficult case.

Time's Stately Cycle as a Key to the Organization of Lyell's Principles

Lyell published eleven editions of the *Principles of Geology* between 1830 and 1872 (see specifications of dates and major changes in the preface to the last edition—Lyell, 1872). Since Lyell regarded his great work as a lifelong source of income, he continually revised, shifted sections and chapters, and experimented with differing formats—much as the author of a modern best-selling text produces new editions, perhaps too frequently, for commercial more than intellectual motives. Lyell's similar behavior has prompted the general misinterpretation of his great work as a textbook in the usual sense. As I argued above (see page 104), it is no such thing; *Principles of Geology* is a brief for a world view—time's stately cycle as the incarnation of rationality.

I believe that all truly seminal works of our intellectual history are coherent arguments for grand visions. Lyell's *Principles* lies squarely in this greatest of all scholarly traditions, and yet, as I argue above, it has usually been read as a work of the most opposite genre conceivable: the textbook, with its pseudo-objectivity and impassive compendium of accepted information.

Grand visions require keys to unbolt their coherence. Often we lose those keys when changing contexts of history bury the motivations of authors in forgotten concerns. Lyell's *Principles* has suffered this fate. The key to its coherence is Lyell's overarching vision of time's stately cycle—the combination of his uniformities of rate and state. But we now view his brief as a textbook because we no longer recognize this thread of unity. The real Lyell has been sacrificed, in part by his own rhetoric, for the cardboard hero of empirical truth. The great thinker, the scientist of vision, the man who struggled so hard to grasp the empirical world as imbued with distinctive meaning, becomes merely a superior scribe.

We can, at least, try to recover Lyell's vision by grasping the *Principles* as an argument, not a compendium, *Time's stately cycle is the thread of coherence, for Lyell's* Principles *is a treatise on method, dedicated to defending this vision in the face of a geological record that requires close interpretation, not literal reading, to yield its secret support.*

I have a personal theory that paradoxes of odd beginnings usually unlock the meaning of great works. Darwin's *Origin of Species* does not announce a revolution in thinking, but starts instead with a disquisition on variation among breeds of pigeons (as Burnet begins by puzzling about floodwaters, and Hutton with the paradox of the soil). When we recognize that Darwin's defense of natural selection is an extended analogy from small-scale events that we can watch and manipulate—artificial selection as practiced by agriculturists, breeders, and fanciers—to invisible events of grander scale in nature, his otherwise eccentric beginning makes sense. Ernst Mayr (1963) began our most important modern book on species and their origin with an empirical list of sibling species,[7] not with general theories or global frameworks. When we grasp Mayr's major aim—to substitute a dynamic view of species as natural populations defined by interbreeding and ecological role for the old taxonomist's idea of dead things that look alike in a museum drawer—we recognize that his choice for starters embodies his book's program: for sibling species are the test case of his vision—perfectly good species by the new criterion, unrecognizable under the old.

Following five historical chapters that tout the factual and moral benefits of nondirection, Lyell begins his substantive brief with three chapters on climate in the northern hemisphere and one on the hypothesis of progression in life's history. Chapter 6 of volume I bears the title "Proofs that the climate of the Northern hemisphere was formerly hotter." And so—mark the oddity—the first substantive chapter in a three-volume brief for time's cycle admits as its central theme the most favorable datum that Lyell's opponents, advocates of a directionally cooling earth, could possibly muster.

Chapter 7, "on the causes of vicissitudes in climate," then argues that varying distributions of land and sea are the most evident and easily ascertainable causes of climatic change. (A vast ocean dotted with a few small islands will bring warmer and more equable climates than a massive continent with little surrounding water at the same latitudes.) Chapter 8 bears an extended title: "Geological proofs that the geographical features of the northern hemisphere, at the period of the deposition of the carboniferous strata, were such as would, according to the theory before explained, give rise to an extremely hot climate." And now we understand the point and the program.

Carboniferous rocks are old. They represent a time of great swamps and lush tropical vegetation; their fossil remains supply

7. Sibling species are morphologically indistinguishable populations as clearly separated by behavior in nature as others more visually distinct.

most of our coal. Superficially, they provide firm support for the directionalist hypothesis of an inherently cooling earth.

Lyell admits the phenomenon—northern-hemisphere coal forests indicate hotter climates during the Carboniferous (I, ch. 6). If an inherent secular cooling caused the subsequent change, then uniformity of state is disproved. But Lyell offers an alternative based on current processes (I, ch. 7). As sea and land change position, climate alters in predictable ways. If the northern hemisphere had become more and more continental since the Carboniferous, then climate would become cooler as a result of fluctuating surfaces, not inexorably cooling interiors. Lyell then tries to demonstrate (I, ch. 8) that cooling climates in the northern hemisphere have been accompanied by increasing continentality since the Carboniferous.

This alternative explanation preserves the crucial uniformity of state. Interiors cooling from an original fireball are irreversible records of time's arrow. But exteriors that cool because continents rise impart no inherent direction to time, and permit no future extension to further frigidity. Continents rise by uplift and fall by erosion in a smooth and nondirectional way through the fullness of time, recording both substantive uniformities of rate and state: "The renovating as well as the destroying causes are unceasingly at work, the repair of land being as constant as its decay, and the deepening of seas keeping pace with the formation of shoals" (I, 473). As continents have emerged since the Carboniferous, they may yield again to ocean in the future—and cooling is but one segment of a reversible cycle. Lyell speaks of a "great year," or "geological cycle," and views falling temperatures in the northern hemisphere since Carboniferous times as the autumn of a geological succession that will see another summer.

The first three substantive chapters are, therefore, one long application of Lyell's method to an apparent (and central) case of disconfirmation. He probes behind appearance to render an admitted phase of cooling, occupying most of geological time and a large part of the earth, as but one arc of a grander circle.

Chapter 9 then applies this reasoning to the greatest problem for any supporter of time's cycle—the apparent increase in life's complexity through time. Lyell again admits the appearance but denies the inherent directionality (see previous section for details). New species always arise perfectly adapted to prevailing climates. If climates become cooler, new species will display increases of complexity suited to these more difficult conditions. Directional trends in life only record an underlying change in climate. If an apparent climatic "arrow" is really the segment of a circle rotating to nowhere, then life will also follow the future arc back—to Professor Ichthyosaurus of times to come.

Having marshaled history to his purposes (chs. 1–5) and dismissed the two most troubling cases of apparent directionality (climate and life, chs. 6–9), Lyell devotes the rest of volume I (chs. 10–26) to a catalogue of modern causes presented as a complete guide to the past. He arranges these chapters (ostensibly about the second methodological uniformity of process) as a subtle defense of his substantive uniformities of rate and state. For he discusses first the aqueous causes (Figure 4.5) that destroy topography (rivers, torrents, springs, currents, and tides) and then the igneous causes (Figure 4.6) that renew (volcanoes and earthquakes), suggesting all the while that both sets operate in continuous balance; neither ever dominates the entire earth, and neither imparts any inherent direction to the character of rocks, landforms, or life.

I shall discuss volumes II and III (Lyell's positive contribution) in the next section, but must briefly note here how they continue and fulfill the plan of a grand, coherent work on life and its nature. We usually remember volume II only for Lyell's refutation of evolution, particularly of Lamarck's theory (which he introduced to England, though only to dismiss). But the eighteen chapters of volume II are designed and sequenced to present a positive view of life's history that will lead to Lyell's greatest achievement, the subject of volume III—a new method for stratigraphic dating based on an unconventional view of life dictated by the uniformities of rate and state.

The focus of Lyell's argument—and the reason for lambasting evolution defined as insensible transition between species—rests upon a view of species as entities, not tendencies; things, not arbitrary segments of a flux. Species arise at particular times in particular regions. They are, if you will, particles with a definite point of origin, an unchanging character during their geological duration, and a clear moment of extinction. Most important, they are particles in a world of time's stately cycle. Their origins and extinctions are not clumped in episodes of mass dying or explosive radiation, but more or less evenly distributed through time, with births balancing deaths to maintain an approximate constancy in

Figure 4.5
A modern example of destruction by erosion. The Grind of the Navir (the breach between the two sections of this sea cliff in the Shetland Islands) is widened every winter by the surge passing between. (From first edition of Lyell's *Principles*.)

Changes of the surface at Fra Ramondo, near Soriano, in Calabria.

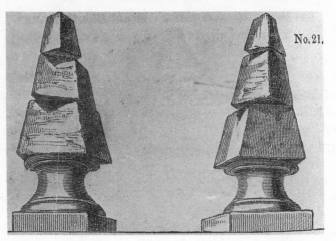

No. 21.

Figure 4.6

Modern examples of construction by earthquakes. Top: the surface at Fra Ramondo in Calabria. Note the breakage of olive trees between up- and downthrusted parts of the hill. Bottom: two obelisks on the facade of the convent of S. Bruno in Stefano del Bosco, Italy. Each is built of three sections, and repeated earthquakes have loosened the blocks and rotated them to different positions. (From first edition of Lyell's *Principles*.)

life's diversity. Their sequential origins display no vector of progress for a positive reason of theory, not a mere claim of observation (see previous section): Lyell espoused perfect adaptation between species and their environments; each species is a mirror of its surroundings. Therefore, any true direction in life's history can only record a corresponding arrow in the physical world. Since the physical world has remained in steady state (volume I), life has also maintained an unchanging complexity and diversity. Species turn over constantly; none alive today graced the Carboniferous coal swamps. But anatomical designs do not accrete or improve.

Volume III, in twenty-six chapters, presents a descriptive account of the earth's actual history, ordered, as a man committed to modern processes might, in a sequence opposite to modern conventions—starting with most recent times (where the work of modern processes can be assessed most readily) and working back to the oldest rocks. Many readers have dismissed volume III as dull and outdated description, but it embodies the central defense of Lyell's game plan. It presents his most important argument for time's stately cycle—for a vision in science is only as good as its application *and its utility*. Volume III, read as the ultimate test of time's cycle, embodies two major purposes. First, the striking vectors of geological history read literally must be interpreted, by Lyell's method of probing behind appearance, as the way that an imperfect record would render time's cycle in preserved evidence. Again and again, we learn (for example) how apparent mass extinctions are periods of nondeposition, and how greater contortion of older rocks records the longer time available for their subsequent modification by constant forces of metamorphism, not the greater vigor of a pristine earth.

Second, Lyell, as a great scientist, understood the cardinal principle of our profession—that utility in action is the ultimate test of an idea's value. Most of his earlier defenses of time's cycle had been rhetorical, verbal, or negative (by showing that directional appearances of the literal record do not disprove a steady state). To crown the success of his brief, Lyell now needed an achievement of substance—something major and practical that time's stately cycle could do to unravel the earth's history. Volume III is therefore, above all, a long illustration of a new method, striking in its originality and brilliant in its difference from conventional paleontology, for dating rocks of the Cenozoic Era (the last 65 million years, since the extinction of dinosaurs) by percentages of molluscan species still living. I shall show in the next section that this novel method flows directly from Lyell's unusual view of time's cycle applied to life's history.

Although (obviously, from early sections of this chapter) Lyell is not my foremost intellectual hero, I can only describe my reading of the first edition of the *Principles* as a thrill, a privilege, and an adventure. As I grasped its brilliant coherence about the vision of time's stately cycle, shivers coursed up and down my spine. Yet that thrill has been foreclosed to most readers. The first edition is difficult to obtain, and many reasons conspire to degrade its coherence through the subsequent editions that most geologists read. For one thing, Lyell extracted almost all of volume III, and placed his discussion of the earth's actual history into another book, the *Elements of Geology* (in later editions, the *Manual of Elementary Geology*)—thus divorcing his primary application from his verbal defense of time's cycle. For another, Lyell strongly muted his commitment to time's cycle when, late in his career, and with both great personal struggle and splendid honesty (see pages 167–173), he finally admitted the progressive character of life's history. Finally, he shifted and tinkered with so many chapters that the original coherence of argument dissipated, and the last editions almost became, after all, a textbook.

Lyell, Historian of Time's Cycle

Lyell's Explication of History

Hutton and Lyell are indissolubly linked in textbook histories as the two heroes of modern geology—Hutton as unheeded prophet,

Lyell as triumphant scribe. Gilluly, Waters, and Woodford, for example, write (1968, 18): "The uniformitarian principle, proposed by James Hutton of Edinburgh in 1785, was popularized in a textbook by the great Scottish geologist Charles Lyell in 1830." We have seen that methodological versions of uniformity were the common property of all scientists, defended by both Hutton and Lyell, but scarcely original with them. (We cannot even label Hutton as a champion of actualism, for he argued that forces of subterranean consolidation were invisible on today's earth, and must be inferred from the character of ancient rocks exposed by uplift.) Hutton and Lyell shared, above all, the controlling vision of time's cycle, the uniformity of state. Even here they differed, for Hutton promoted a sequential view, and held that periods of uplift might be global and catastrophic, while all stages of Lyell's cycle operate locally and simultaneously, giving the earth a timeless steadiness through all its dynamic churning. An observer might visit Hutton's earth and see only the quietude of subterranean deposition, while another visitor, a million years later, might find a planet convulsed by uplift. The pieces of Lyell's globe shift constantly, but all processes are always working somewhere—at about the same intensity and amount.

I do not, however, view Lyell's union of rate and state (time's stately cycle), and Hutton's more catastrophic notion of uplift, as their most important difference. We need to recapture the dichotomy of their day—time's arrow versus time's cycle—to grasp their deeper divergence in different attitudes toward the meaning of time's cycle.

Hutton carried out his strict version of the Newtonian program so completely that his view of our planet became idiosyncratic—to a point where he actually denied the subject that students of the earth have always advanced as their fundamental motivation: history itself, defined as a sequence of particular events in time. Temporal distinction has no meaning in Hutton's world, and he never used the language of historical uniqueness to describe the earth. The corresponding events of each cycle are so alike that we can scarcely know (or care) where we are in a series that displays no vestige of a beginning, no prospect of an end. (I also noted that John Playfair, Hutton's Boswell, did not follow Hutton's idiosyncrasy, but used the ordinary language of historical narrative in his explication of Hutton's theories. Therefore, since most scientists know Hutton only through Playfair, this special character of Hutton's system has been lost.)

Lyell shared Hutton's commitment to time's cycle, but not his ahistorical vision, for reasons both personal and chronological. The fifty years separating Hutton and Lyell had witnessed a transformation in practice among British geologists. Hutton crowned a tradition of general system-building, or "theories of the earth." The next generation had abjured this procedure as premature and harmful speculation. The nascent science of geology needed hard data from the field, not fatuous, overarching theories. The eschewing of "interpretation," and restriction of discussion to facts alone, was (however impossible the ideal), actually written into the procedures of the Geological Society of London, founded in 1807. As a primary approach to field evidence, embraced for its plethora of exciting results, the Geological Society adopted the stratigraphic research program. The primary task of geology must be defined as unraveling the sequence of actual events in time, using the key to history that had just been developed to the point of general utility by Cuvier and William Smith—the distinctively changing suite of fossils through time.

Lyell was a willing child of this transformation in procedure. He was a historian, and the primary data of history are descriptions of sequential events, each viewed as unique (lest any lack of distinction blur utility as indicator of a particular moment). Lyell could scarcely embrace Hutton's antihistorical viewpoint. But how could a committed historian defend *and use* time's cycle? And how could nondirectionalism aid the stratigraphic research program as a tool for unraveling historical sequences?

The very first line of Lyell's *Principles* stakes out his difference from Hutton's ahistorical vision: "Geology is the science which investigates the successive changes that have taken place in the organic and inorganic kingdoms of nature" (I, 1). The statement seems so innocuous, but search far and wide in Hutton and you will never find its like; major upheavals in thought often sneak past our gaze because their later success makes them seem so obvious.

Lyell's first words show a profound understanding of both the meaning and the joy of history. He begins by acknowledging the distinctive character of historical inquiry—the explanation of present phenomena as contingent results of a past that might have been different, not as predictable products of nature's laws. The original historical prod may be tiny and forgotten, but results cascade to a magnitude that often seems to belie their origin:

> We often discover with surprise, on looking back into the chronicles of nations, how the fortune of some battle has influenced the fate of millions of our contemporaries, when it has long been forgotten by the mass of the population. With this remote event we may find inseparably connected the geographical boundaries of a great state, the language now spoken by the inhabitants, their peculiar manners, laws, and religious opinions. But far more astonishing and unexpected are the connexions brought to light, when we carry back our researches into the history of nature. (I, 2)

A static analysis of current function may yield some insight, but consider the expansion provided by historical context:

> A comparative anatomist may derive some accession of knowledge from the bare inspection of the remains of an extinct quadruped, but the relic throws much greater light upon his own science, when he is informed to what relative era it belonged, what plants and animals were its contemporaries, in what degree of latitude it once existed, and other historical details. (I, 3)

Recognizing the importance of taxonomy, Lyell sought to rank geology properly among the sciences. He refused to follow several predecessors because they had placed geology with physical sciences based on laws of nature that impart no distinctive historical character to present phenomena. Thus, Werner had viewed geology as "a subordinate department of mineralogy" (I, 4), and Desmarest as a branch of physical geography. But minerals owe their properties to chemical composition, landforms to physical agents of uplift and erosion. Neither discipline acknowledges the irreducibly historical character of geological phenomena. Another proposed union with cosmogony must also be rejected; for, while geology requires a place among sciences of history, it must be defined as an empirical study of preserved records, and not linked with mental excursions about the origins of things.

As a direct study of history, geology owes its stunning success to a great transformation of practice, then just a generation old—the stratigraphic research program, using fossils as the key to ordering by age: "In recent times, we may attribute our rapid progress chiefly to the careful determination of the order of succession in mineral masses, by means of their different organic contents, and their regular superposition" (I, 30).

In a forceful passage, Lyell identifies the distinctive method of history—we need not ape the quantitative procedures of physical science, but must celebrate the power of what others with less understanding might deem a humdrum occupation, the ordering of events in time.

> By the geometer were measured the regions of space, and the relative distances of the heavenly bodies—by the geologist myriads of ages were reckoned, not by arithmetical computation, but by a train of physical events—a succession of phenomena in the animate and inanimate worlds—signs which convey to our minds more definite ideas than figures can do, of the immensity of time.

Physics used its techniques to expand space; we have employed ours to enlarge time—results scarcely matched for significance in the history of thought, but won by different methods.

Lyell recognized that real history must be a "succession of phenomena." Precise cyclical recurrence would blot the distinctive character of historical moments. Lyell ridicules the old cyclical theories of Egypt and Greece, the *ewige Wiederkehr* or eternal return, but this passage might apply just as well to Hutton's ahistorical vision:

> For they compared the course of events on our globe to astronomical cycles . . . They taught that on the earth, as well as in the heavens, the same identical phenomena recurred again and again in a perpetual vicissitude. The same individual men were doomed to be re-born, and to perform the same actions as before; the same arts were to be invented, and the same cities built and destroyed. The Argonautic expedition was destined to sail again with the same heroes, and Achilles with his Myrmidons, to renew the combat before the walls of Troy.   (I, 156–157)

Finally, we must never lose the simple and unvarnished joy of discovering a past that had disappeared from view: "Meanwhile the charm of first discovery is our own, and as we explore this magnificent field of inquiry, the sentiment of a great historian of our times [Niebuhr, author of the *History of Rome*] may continually be present to our minds, that 'he who calls what has vanished back again into being, enjoys a bliss like that of creating'"   (I, 74).

Dating the Tertiary by Time's Stately Cycle

Since preservation improves with recency in the geological record, we might anticipate that the youngest rocks would be most easy to resolve by the stratigraphic research program. Old (Paleozoic) rocks are often twisted and metamorphosed, their fossils distorted, pulverized, or entirely leached away. Geologists did struggle with the Paleozoic, and its resolution was a triumph and test of ultimate utility for the stratigraphic research program (see Rudwick's brilliant account of the Devonian controversy, 1985).

Tertiary strata of the "age of mammals" (all but the tail end of the last 65 million years by modern reckoning) should have succumbed first to resolution, as a test case for the new techniques. Paradoxically, by contingent bad fortune of the particular history of Tertiary times in Europe, these youngest rocks provoked puzzle rather than resolution. The middle or so-called Secondary strata (including what we now call Mesozoic and the top part of the older Paleozoic) yielded first to the new methods. These rocks are ordered throughout Western Europe in the nearly ideal configurations of textbook dreams—as extensive sheets of minimally distorted, flat-lying or gently tilting beds, easily traced over large areas. For example, the distinctive "chalk" (forming, for example, the White Cliffs of Dover), top layer of the Secondary, blankets this region with little complexity in deposition or later distortion—you really can't miss it, as the saying goes.

By contrast, the younger Tertiary strata are deposited as a complex patchwork in isolated basins—the bane of any stratigrapher's existence (Figure 4.7). Geologists work by correlation and super-position—fancy words for the obvious techniques of ascertaining what beds lie atop others (superposition), and then tracing this order from place to place (correlation). But if strata are clumps

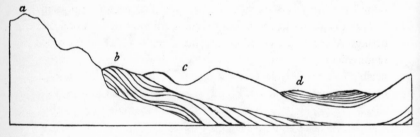

a, Primary rocks.
b, Older secondary formations.   c, Chalk.
d, Tertiary formation.

Figure 4.7

An illustration of the problems faced by geologists in unraveling Tertiary stratigraphy. Tertiary rocks of Europe tend to occur in small and isolated basins (as in d above), making correlation difficult. (From first edition of Lyell's *Principles*.)

rather than sheets, then we cannot unravel them by superposition. (Many Tertiary strata are, for example, impersistent stream channels rather than broad sheets of shallow oceanic sediments so common in the Mesozoic.) And if strata are confined to isolated local basins, then we cannot correlate them easily from place to place.

Since Tertiary times had been marked by increasing continentality in Europe (recall Lyell's second substantive argument; see pages 139–141), marine sediments were deposited in shifting, isolated embayments, not as favored broad sheets. Thus, Tertiary strata were a challenge to the stratigraphic program, not its premier example, as logic (without history's peculiarities) might have dictated. They were also something of an embarrassment, since a good technique should snare its potentially easiest reward without difficulty. Lyell therefore decided to bag the Tertiary with a different method, based on his distinctive vision of time's cycle. Success would crown his abstract vision as weighed in the empirical balance and found triumphant.

In the face of such difficult stratigraphy, fossil remains would unlock the Tertiary sequence. The stratigraphic research program had congealed about the paleontological criterion of temporal ordering. A chronometer of history has one, and only one, rigid requirement—something must be found that changes in a recognizable and irreversible way through time, so that each historical moment bears a distinctive signature. Geologists had long appreciated this principle in the abstract, but had not found a workable criterion. Werner and the Neptunians had tried to use rocks themselves, arguing that a distinctive suite of compositions and densities had precipitated in temporal sequence from a universal ocean. This idea was sound in logic, but didn't work in practice because the earth's strata were not deposited in order of density from one ocean in one great era of precipitation. Moreover, rocks are simple physical objects formed by chemical laws and, as such, do not bear distinctive temporal signatures. Quartz is quartz—conjoined tetrahedra with a silicon ion in the center, surrounded by four oxygen ions, each shared with a neighboring tetrahedron. So it was in the beginning, and is now, and ever shall be so long as nature's laws prevail. Cambrian quartz is no different from Pleistocene quartz.

But life is complex enough to change through a series of unrepeated states. Today we attribute this irreversible sequence to the workings of evolution, but the fact of uniqueness may stand prior to any theory invoked to resolve it. The fossil criterion became the Rosetta stone for the stratigraphic research program, but few early consumers accepted evolution as the reason behind distinctive temporal stages of the fossil record. In Lyell's time, the fact of temporal distinction stood as an unexplained but crucial tool. Lyell himself had always professed agnosticism about the reasons, stating that he simply did not know whether new species arose by God's direct will, or by the operation of unknown secondary causes—though he did profess confidence that they arose in perfect balance with their environments.

In Lyell's time, the problem of unresolved Tertiary stratigraphy centered upon a proper use of fossils to "zone" strata—that is, to establish a worldwide sequence of temporally ordered stages within this long and previously undivided segment of the earth's history. In 1830, most stratigraphers were progressionists. They believed that life had improved throughout the Tertiary and that, as a rough guide at least, we might judge the relative age of Tertiary strata by the level of development displayed in their fossils. All paleontologists understood that progress was neither sufficiently linear nor unambiguous enough to employ as an actual measuring rod. In practice, a progressionist might use similarity to living forms, rather than some unattainable measure of relative perfection in biomechanical design. Moreover, he would search for a series of guide fossils—easily identifiable creatures of short and distinct geological range—to zone the Tertiary. And he would focus more upon their uniqueness and restriction to a small stretch of time than upon their supposed levels of relative complexity. Still, however far practice diverged from theory, a commitment to progressionism still channeled the actual work of stratigraphy into a search for temporal sequences of guide fossils arrayed as a ladder of improvement.

Lyell's world view did not permit him to work by the usual methods of his profession. Life participated fully in the dynamic steady state of time's cycle. The fossil record displayed no vector of progress, and its sequence of species could not be ordered by any criterion of advance. To put it as baldly as possible, life (as a totality) was always just about the same—with balance maintained both in number of species and relative proportions of different groups. How, then, could a Lyellian find any paleontological criterion for dating rocks—and, if not, how could he participate in the central and guiding concern of his profession?

Had Lyell been a strict Huttonian, he would have found no exit from this dilemma. He would have been mired in an ahistorical outlook that viewed each event as so similar to its corresponding stage in the previous cycle that no criterion of history could be established. But Lyell, as a premier practicing geologist of his day, was a committed historian. He accepted the uniqueness of events, and used this principle to extract a mark of history from time's cycle.

To borrow Lyell's own favored technique of metaphorical illustration, we may depict all the earth's species at any one time as a fixed number of beans in a bag—for species are particles in Lyell's vision. We begin a five-day experiment. The bag contains a thousand beans, and it will always hold this number. New beans are entering at a fixed and constant schedule, say one every two minutes. But the bag can only hold a thousand beans, so each time a new one enters, the beanmaster reaches in and pulls an old one out at random.

One more crucial step completes the isomorphism with Lyell's view of life. The beans are not identical; each is a distinct historical object. Let us say that each bears a unique brand in its lower right-hand corner (if beans may be construed as possessing such a thing). We can tell unambiguously which bean is which—but, and here's the rub, these distinct brands include *absolutely no signature of time whatever*. The beans are not color-coded by day of entry into the bag, or marked with the geometry of their time of origin. In other words, we can recognize each bean as an distinctive object, but we have no clue (from form or color) about its age or time of entry into the bag.

This system corresponds point for point with Lyell's vision of the fossil record in a world of time's stately cycle. The five days are the broad eras of geological time (few in number); the brands are marks of historical uniqueness, but (please note) not of progress, for each bean is distinct and all are equivalent in merit. The entrance of a bean every two minutes marks the stately uniformity of rate; the random removal of an old bean at each entrance maintains the steady state of diversity.

The grand beanmaster now sets us a problem. He took an x-ray of the bag every six hours during the last day, but he forgot to mark the times on his negatives, and he wants us to arrange the four photos (for midnight, six A.M., noon, and six P.M.) in proper temporal order. He is also willing to give us the bag as now constituted at day's end. How can we proceed?

Lyell and his student Simplicio consider the problem. Simplicio, ever in search of the easy way, suggests that they look for a crucial bean in each photo. But Lyell responds that no such object can exist. The uniqueness of each bean is, perversely, absolutely no guide to its age. Lyell castigates Simplicio for laziness, and argues that the problem can only be approached statistically.

Fortunately, the wily beanmaster has provided one criterion that can resolve this dilemma of history under time's cycle—he has given us the bag in its present state. Consider, Lyell advises, how the beanmaster proceeded on the last day. Every two minutes, or 720 times during the day, he put in a new bean and removed one from the bag at random. We now open the bag. It is dominated, we reason, by beans added during the last day—not all 720, of course, because some have been removed by luck of the (with)draw. But, Simplicio complains as he begins to catch on, we can't tell from the signatures which beans represent the last day's additions, for a signature contains no information *in se* about time.

Lyell then proposes his statistical criterion. We cannot know when any particular bean entered the bag, but we can make a list of all signatures in the bag as now constituted. We can then study the beanmaster's four photos and tabulate the 1000 signatures in each. The longer any bean is in the bag, the greater its chance of removal (since beans are pulled out at random as new ones enter). Thus, the more recently any bean entered, the greater the chance that it still resides in the bag. Lyell exclaims triumphantly that we need only tabulate, for each photo of the bag at a previous time, the percentage of beans that remain in the bag at day's end. The higher the proportion of current beans, the younger the photo.

What other criterion could we use? No bean betrays its age, but we do have the present bag, and can tell time by a gradual and continuous approach to current composition. The present bag is not better or more distinctive than any other; it only possesses the virtue of precise location in time—and we may therefore compare the similarity of other bags with it.

Lyell dated the Tertiary (last "day" of just a few) in this manner precisely. He proposed a statistical measure based on the relative percentage of living species of mollusks. (He used mollusks because they are numerous and distinct in Tertiary strata, and because he could pay his French colleague Deshayes to compile faunal lists for all major sections of the European Tertiary, based on the consistent criterion of Deshayes's personal taxonomic expertise.) Such a statistical criterion cannot yield overly fine distinctions, since several random factors operate (not just in removal, as in our bean experiment, but also, for the complexities of reality, because species don't arise at equally spaced intervals, and because total numbers of species are not truly constant). Thus, Lyell split the Tertiary into four subdivisions (as the beanmaster took four photos)—named, in order, Eocene. Miocene, older Pliocene, and newer Pliocene (Figures 4.8 and 4.9), and defined as bearing about 3 percent of living species (Eocene), about 20 percent (Miocene), more than a third and often more than a half (older Pliocene), and about 90 percent (newer Pliocene). (Readers who have been forced to memorize the geological time scale will recognize that our modern system retains Lyell's names, with a few additions in the same mold—Paleocene,

Figure 4.8
Eocene mollusk fossils used by Lyell in his statistical method for zoning the Tertiary. (Plate 3, volume 3 of the first edition of Lyell's *Principles*.)

Figure 4.9

Miocene mollusk fossils used by Lyell to zone the Tertiary. (Plate 2, volume 3 of Lyell's *Principles*.) Note, by comparing figures 4.8 and 4.9, the key precepts of Lyell's method, based on the assumption of time's stately cycle. The later Miocene fossils are in no way "better" versions of their forebears. They are simply different as signs of history's passage.

Oligocene, and Pleistocene—to express the finer divisions permitted by increasing knowledge.) Lyell writes, citing a metaphor strictly comparable with our beanbag:

> This increase of existing species, and gradual disappearance of the extinct, as we trace the series of formations from the older to the newer, is strictly analogous, as we before observed, to the fluctuations of a population such as might be recorded at successive periods, from the time when the oldest of the individuals now living was born to the present moment.

This simple description of Lyell's method cannot capture the brilliance and radical character of his concept. Consider just three points:

First, Lyell proposed this numerical method based on a sophisticated model of random processes at a time when such statistical thinking was in its infancy. Most of us still need metaphors to grasp it today, after a century of success for this powerful procedure.

Second, Lyell's method flies in the face of all paleontological convention during his time. Most stratigraphers denied both substantive uniformities of rate and state—and each denial led to a method abjured by Lyell. I have already discussed how progressionism (nonuniformity of state) led most stratigraphers to search for key fossils that might mark time by their anatomical complexity—a method contrary to Lyell's statistical approach toward entire faunas.

Lyell's opponents also rejected uniformity of rate by viewing the fossil record as punctuated by mass extinctions and rapid subsequent radiations of new species. This concept of life's history led to a different practice in dating—the search for distinctive suites of species to mark each epoch of time. Such a procedure makes no sense in Lyell's world. His species are independent particles evenly spaced in time; they do not enter and leave the geological scene in concert. Distinct epochs are an illusion of our imperfect record; we can only capture moments with statistical measures of a smooth and continuous flow:

We are apprehensive lest zoological periods in Geology, like artificial divisions in other branches of Natural History, should acquire too much importance, from being supposed to be founded on some great interruptions in the regular series of events in the organic world, whereas, like the genera and orders in zoology and botany, we ought to regard them as invented for the convenience of systematic arrangement, always expecting to discover intermediate gradations between the boundary lines that we have first drawn.   (III, 57)

Third—and this is harder to put in words—Lyell's method is quirky and fascinating. It makes you think. It stands against all traditions of the field, from his day to our own. Paleontologists are devoted to specifics. Professionals become experts on particular groups at particular times; we receive advanced training by apprenticeship to authorities, and we spend years learning the taxonomic details of our chosen group. We expect to resolve stratigraphic problems by using this expertise—to identify this stretch of time because Joe the brachiopod lives there, and that interval because Jill the bryozoan inhabits its strata.

"Charles Lyell's dream of a statistical paleontology" (the apt title of Rudwick's fine analysis, 1978) stands against this tradition of particulars. It applies to the quirky world of history the generality of an abstract process, regular as the ticking of a clock (albeit with random fluctuations). Such marriages of dissimilar partners (methods of one domain with alien particulars of another) are often among the most fruitful of intellectual unions.

This understanding of Lyell's ingenious method for dating the Tertiary also permits us to grasp the rationale and organization behind the last two volumes of the *Principles of Geology*, for both are centered upon Lyell's attempt to apply time's cycle as a working method of history. We may view volume II, with some simplification, as a long defense of "species as particles" in a world of time's stately cycle—setting up, if you will, the metaphor of the beanbag. With this key, we understand Lyell's true design and do not misread him as an old fuddy-duddy castigating evolution even before Darwin tried to make the idea popular.

Chapters 1–4 of volume II, Lyell's attack on Lamarck and the entire concept of evolution, argue that species are particles, not tendencies or arbitrary segments of a continuous flux. Species are beans in nature's bag. In the closing words of chapter 4: "It appears that species have a real existence in nature, and that each was endowed, at the time of its creation, with the attributes and organization by which it is now distinguished" (II, 65).

Chapters 5–8, on geographic distribution, claim that species arise at particular places, in foci of origin. Again, they are not general tendencies, but particular things—beans that enter the bag as unique items at definable moments. Lyell writes that "single stocks only of each animal and plant are originally created, and that individuals of new species do not suddenly start up in many different places at once" (II, 80).

Chapters 9–10 then discuss the principle of perfect fit to prevailing environment, the branding of each bean with its distinctive [8] signature: "the fluctuations of the animate and inanimate creation should be in perfect harmony with each other" (II, 159). Finally, chapter 11 argues that introduction of new species compensates for gradual loss of old forms—the beanbag remains in dynamic balance, always full but changing in composition: "the hypothesis of the gradual extinction of certain animals and plants and the successive introduction of new species" (III, 30).

Volume III is an excursion through geological time—an application of Lyell's methodology to the earth's actual history. But

8. Lest this claim seem inconsistent with Lyell's reverie about returning ichthyosaurs, I note that he chose his words with consummate care. The famous passage, caricatured by De la Beche, states: "then might those genera of animals return," not "then might those species . . ." The difference is crucial, not trivial. Genera, to Lyell, are arbitrary names for anatomical designs; species are unique particulars. The returning summer of the great year might inspire the origin of a creature sufficiently like Jurassic ichthyosaurs that taxonomists would place it in the same genus. But the returning ichthyosaur would be a new bean, with distinctive characters marking it as a unique species.

nineteen of its twenty-six chapters chronicle the Tertiary and most others discuss Tertiary problems prominently. The volume ends with a sixty-page appendix, reproducing *in toto* Deshayes's charts for the duration of Tertiary mollusks and the percentage of living species in each stratigraphic unit. We have no trouble detecting Lyell's main interest; for this is no impartial text, allotting space in proportion to time or preserved strata.

Most working geologists could tell you that Lyell named the epochs of the Tertiary. They know this as a curious little fact, proving that the apostle of uniformitarianism also did some field-work. If we could only learn to grasp the intimate—indeed necessary—connection of this achievement with his vision of time's cycle, then we would understand the power of Lyell's system. Lyell broke through the sterility of Hutton's ahistorical view, and showed that the vision of time's stately cycle could serve as a research tool for geology's basic activity, the ordering of events in time. Lyell's system works because we inhabit a world of history—by the primal criterion of uniqueness, based on temporal context, for each phenomenon. Charles Lyell was the *historian* of time's cycle.

The Partial Unraveling of Lyell's World View

Retreating from the Uniformity of State, or Why Lyell Became an Evolutionist

Mountains arise and erode through time—"the seas go in and the seas go out," as the old geologists' motto proclaims. Uniformity of state might well describe physical history. But Lyell's extension of time's cycle to the history of life had always seemed implausible to most colleagues—especially in the light of human origins at the very summit of time's mountain. Lyell had provided a rationale for nonprogressionism in life's history (see pages 137–142), but his arguments were shaky on both theoretical and empirical grounds. Thus, when Lyell, late in his career, finally surrendered his uncompromising commitment to time's cycle, he capitulated by admitting, albeit with great reluctance, that apparent progress in life's history was also a reality after all.

Lyell held firm for more than twenty years, from the first edition of the *Principles* in 1830 to his last defense of nonprogression in an anniversary address as president of the Geological Society of London in 1851. But twenty years of exploration had uncovered no Paleozoic mammals, and his old argument—that we had no right to expect any while our knowledge of Paleozoic times rested upon just a few oceanic sediments of limited geographic extent—became less and less defensible as studies of Paleozoic geology spread into eastern Europe and North America. Lyell began to waver, and eventually, during a painful process extending through the 1850s, he surrendered.

While Lyell believed that no human remains, or even artifacts, graced the geological record, he could view *Homo sapiens* as God's addition of the last moment. But as undoubted artifacts were unearthed from the youngest strata, Lyell could no longer deny that human origin had been an event in the ordinary course of nature. How could he then deny progress as a guiding principle? Thus, when Lyell gathered his material on human history into a separate volume in 1862 (*On the Geological Evidences of the Antiquity of Man*), he wrote that progress in life's history was "an indispensable hypothesis . . . [which] will never be overthrown."

As for the *Principles,* Lyell had published his ninth edition in 1853, last to defend the strict version of time's cycle. He then waited thirteen years, far longer than the time to any previous revision, to bring forth the tenth edition in 1866—the first to announce his retreat. I have no doubt that this long interval records his growing doubt and confusion, his unwillingness to commit himself once again in print until he had resolved this crucial dilemma. The eleventh edition, last of Lyell's lifetime, appeared in 1872, with only minor revisions from his key capitulation in 1866.

Chapter 9 of this last edition still treats the same subject—"theory of the progressive development of organic life"—but this time Lyell assents. In the summary of his closing paragraph, Lyell finally untangles the conflation of methodological and substantive uniformities that had fueled his rhetoric for forty years—and admits that a scientist can accept progression in life's history while holding firm to the uniformities of law and process:

> But his reliance need not be shaken in the unvarying constancy of the laws of nature [uniformity of law], or in his power of reasoning from the present to the past in regard to the changes of the terrestrial system [uniformity of process], whether in the organic or inorganic world, provided that he does not deny, in the organic world at least, the possibility of a law of evolution and progress.   (1872, I, 171)

Lyell tries to mitigate the meaning of his conversion, depicting it as a minor shift imposed by evidence, and characterizing his former denial of progress as simple skepticism based on insufficient data, but we can scarcely credit this minimization of change. For, by admitting progress in life's history, Lyell surrendered both his vision and all its sequelae, including his dream of a statistical paleontology.

> The dates of the successive appearance of certain classes, orders, and genera, those of higher organization always characterizing rocks newer in the series, have often been mis-stated, and the detection of chronological errors has engendered doubts as to the soundness of the theory of progression. In these doubts I myself indulged freely in former editions of this work. But after numerous corrections have been made as to the date of the earliest signs of life on the globe, and the periods when more highly organized beings, whether animal or vegetable, first entered on the stage, the original theory [progressionism, not uniformity of state] may be defended in a form but slightly modified.   (1872, I, 145)

I believe that Lyell's retreat and surrender usually receive a backward interpretation—another unfortunate result of our anachronistic tendency to impose Darwin's theory of natural selection upon older debates and then to interpret them as part of a great dichotomy between evolution and creation. This reading holds that evolution provoked Lyell's reconsideration—and that his personal linkage of transmutation with progress[9] forced his reassessment once Darwin had convinced him to accept evolution. Lyell was, of course, one of Darwin's principal friends and confidants, a party to the "delicate arrangement" that printed Wallace's separate discovery of natural selection along with an earlier unpublished manuscript of Darwin's, thus affirming Darwin's priority. Charles Darwin was instrumental in provoking Lyell's assent to evolution, and this reversal also enters the 1866 revision of *Principles*.

But a remarkable series of documents—seven private journals "on the species question" compiled by Lyell between 1855 and 1861 and first published by L. G. Wilson in 1970—forces us to reverse this conventional argument toward a new interpretation that makes more sense in terms of human psychology, as usually expressed.

The notebooks record that Darwin first broached his theory to Lyell during a visit at Down in April 1856. (Lyell knew, of course, that Darwin had been working on "the species problem," and that he accepted the common heresy of evolution, but Darwin had not previously revealed his mechanism of natural selection to Lyell.) The journals also show that Lyell was already obsessed with doubt about his linchpin of nonprogressionism in life's history. All accumulating evidence tended to refute his conviction, particularly the discovery of human artifacts in young sediments. He had already, before Darwin's revelation, come to the most reluctant conclusion

---

9. This, in itself, underscores a curious point. No logical necessity can extract an implication of progress from the fact of evolution. Darwin himself had maintained a very ambiguous attitude toward the idea of progress, accepting it provisionally as a feature of parts of the fossil record, but denying that the theory of natural selection—a statement about adaptation to shifting local environments—required organic advance. Nonetheless, many evolutionists have always viewed the concepts of progress and transmutation as necessarily connected, and Lyell, for whatever reason, certainly adopted this view. Therefore, for him, a decision to embrace evolution also entailed progress as a fact: "The progression theory, which accounts for man being improved out of an anthropomorphous species, is natural the moment we embrace the Lamarckian view" (in Wilson, 1970, 59).

of his professional life—that he would probably have to abandon this anchor of his central vision.

What does a man do in the face of such sadness? I suggest that he usually attempts to cut his losses and to beat a minimal retreat. Evolution served Lyell as touchstone for this minimal retreat. Lyell didn't accept evolution because Darwin persuaded him, or because he found the theory of natural selection so powerful; he finally embraced transmutation because it permitted him to preserve all other meanings of uniformity, once accumulating evidence had reluctantly forced him to accept the fact of progress in life's history.

If Darwin impressed Lyell with natural selection, the notebooks record no hint. These private jottings are distinguished by Lyell's nearly total lack of interest in mechanisms of evolutionary change—a decidedly peculiar attitude if Lyell wavered because Darwin's theory had convinced him. The entries record a few passages of criticism, for Lyell never accepted natural selection, much to Darwin's disappointment. I particularly like Lyell's Hindu metaphor, so well expressing the classical objection that natural selection may act as an executioner of the unfit, but cannot create the fit: "If we take the three attributes of the deity of the Hindoo Triad, the Creator, Brahma, the preserver or sustainer, Vishnu, and the destroyer, Siva, Natural Selection will be a combination of the two last but without the first, or the creative power, we cannot conceive the others having any function" (in Wilson, 1970, 369).

The notebooks instead, with almost obsessive repetitiveness, record Lyell's struggle about progress in life's history—particularly his supreme reluctance to place human origins into nature's ordinary course. Yet, when finally forced to admit both the fact of progress and the inclusion of humans in life's standard sequence, what strategies could Lyell adopt to explain his retreat? He imagined only two alternatives—he could either accept the progressionist creed *in toto*, and admit both extraneous laws of progress and (most distasteful of all) perhaps even periods of mass extinction with subsequent recreation at higher levels of complexity; or he could explain the same phenomenon of progress as a consequence of evolution. A key passage reveals that Lyell's distress centered on the fact of progress, and that he viewed evolution as a more acceptable explanation for life's advance than old-fashioned progressionism in its unvarnished form:

> There is but little difference between the out and out progressionist and Lamarck, for in the one case some unknown *modus operandi* called creation is introduced and admitted to be governed by a law causing progressive development and by the other an extension or multiplication by Time of the variety-making power is adopted instead of the unknown process called Creation. It is the theory of a regular series of progressively improved beings ending with Man as part of the same, which is the truly startling conclusion destined, if established, to overturn and subvert received theological dogmas and philosophical reveries quite as much as Transmutation . . . There seems less to choose between the rival hypotheses [evolution and progressionism] than is usually imagined.   (in Wilson, 1970, 222–223)

I regard this last statement (repeated with little variation many times throughout the journals) as the key to Lyell's conversion. He does not accept evolution because facts proclaim it—for he finds little to choose between evolution and progressionism as an explanation for the phenomenon of improvement, now reluctantly admitted. Why, then, prefer evolution?

Lyell's answer seems clear in the journals: evolution is the fallback position of minimal retreat from the rest of uniformity, once life's progress be admitted. If progressionism be embraced, the uniformity of rate will be threatened as well because mass extinction had long been the foundation of progressionist mechanics. Even the uniformity of law might be challenged, if an essentially mysterious process of creation be advocated as the cause of origin (remember that Lyell had never been a creationist, but agnostic about modes of origin for new species). And what about the uniformity of process? Since the creative power operates intermittently and has never been observed on our planet, how can we learn about it by actualist procedures?

But with evolution, Lyell could shore up his defenses and relinquish only one of the uniformities—his beloved time's cycle to be sure, but better one room than the entire edifice. With evolution, he could hold firm to uniformity of rate, especially with Darwin's congenial commitment to such a strict form of *natura non facit saltum* (nature does not make leaps). He could also continue to embrace both the uniformity of law, for evolution "has the advantage of introducing a known general Law, instead of a perpetual intervention of the First Cause" (in Wilson, 1970, 106)—and actualism, for Darwin insisted that small-scale changes produced by breeders and planters were, by extension, the stuff of all evolutionary change.

In short, Lyell accepted evolution in order to preserve his other three uniformities, thereby to retain as much of his uniformitarian vision as possible, when facts of the fossil record finally compelled his reluctant allegiance to progression in life's history. Although I interpret Lyell's embrace of evolution as the most conservative intellectual option available to him, we must not diminish the pain and trouble of mind that it provoked. Consider this remarkable passage, with its resplendent affirmation of both human intellect and basic honesty before the world's complexity:

Species are abstractions, not realities—are like genera. Individuals are the only realities. Nature neither makes nor breaks molds—all is plastic, unfixed, transitional, progressive, or retrograde. There is only one great resource to fall back upon, a reliance that all is for the best, trust in God, a belief that truth is the highest aim, that if it destroys some idols it is better that they should disappear, that the intelligent ruler of the universe has given us this great volume as a privilege, that its interpretation is elevating. (in Wilson, 1970, 121)

The Uniformity of Rate

Since Lyell finally abandoned time's cycle for life's history, this linchpin of his original vision has dropped from sight; few practicing geologists are aware that Lyell ever espoused uniformity of state, and they do not understand the theory of their founding father because they do not recognize its keystone.

But gradualism, or uniformity of rate, experienced a different fate. If anything, Lyell strengthened his commitment to this other substantive uniformity by accepting evolution in Darwin's gradualistic version. Uniformity of rate has therefore persisted to our present day, not always embraced by geologists, but understood as Lyell's vision. Unfortunately, Lyell's trope of rhetoric has also descended in unmodified form—his conflation of method and substance. For more than a century, many geologists have been stifled—the range of their hypotheses falsely channeled and restricted—by a belief that proper method includes an *a priori* commitment to gradual change, and by a preference for explaining phenomena of large scale as the concatenation of innumerable tiny changes.

Lyell's own attempts to base a research program on uniformity of rate failed when his statistical method for zoning the Tertiary foundered upon inconsistent criteria among experts for the designation of species (Rudwick, 1978) and, especially, when he could not extend his method beyond the Tertiary to formulate a general practice rooted in time's stately cycle. If uniformity of rate really applied to the introduction of species—if life's beanmaster introduced and removed these basic units at a stochastically constant rate—then Lyell could have extended his method to the abyss of time. Few modern species could be found in Eocene rocks (defined as 3 percent of modern forms), and none in earlier strata. But, in principle, new baselines could be established to push Lyell's method further back—one might, for example, tabulate the list of Eocene species and then zone the Secondary strata by percentage of species still living in the Eocene.

Lyell did envision just such a procedure. With courage, he designated a difficult, and potentially disconfirming, case as a potential test for his statistical paleontology based on time's stately cycle. He noted a disabling impediment for any scheme of dating Secondary strata by percentage of species still living in Eocene times (first division of the Tertiary). He studied the Maastricht beds, top units of the Secondary, and noted that they contained not a single species also found in Eocene strata. But what could produce such a discordance, for Eocene rocks lay directly atop the Maastricht? In Lyell's world of gradualism, this peculiar circumstance could bear only one interpretation—an immense period of nondeposition, longer than the entire Tertiary, must separate Maastricht and Eocene beds. The beanmaster's cycle had run an entire course during this interval of no preserved evidence:

> There appears, then, to be a greater chasm between the organic remains of the Eocene and Maastricht beds, than between the Eocene and Recent strata; for there are some living shells in the Eocene formations, while there are no Eocene fossils in the newest secondary group. It is not improbable that a greater interval of time may be indicated by this greater dissimilarity in fossil remains ... We may, perhaps, hereafter detect an equal, or even greater series, intermediate between the Maastricht beds and the Eocene strata [than between Eocene and Recent].   (III, 328)

A gutsy prediction required by time's stately cycle, but wrong as we now know. Lyell's catastrophist opponents had long advocated an obvious alternative: no huge gap in time separates Maastricht and Eocene beds; rather, a catastrophic episode of mass extinction marked the end of Secondary times—and this great dying, rather than an immensity of interpolated time based on no evidence, explains the discordance of faunas. We now know that the catastrophists were right. The Cretaceous-Tertiary transition (as we now call it) stands among the five great episodes of mass extinction that have punctuated the history of life. It removed the dinosaurs and their kin, along with some 50 percent of all marine species.

Lyell's gradualism has acted as a set of blinders, channeling hypotheses in one direction among a wide range of plausible alternatives. Its restrictive effects have been particularly severe for those geologists who succumb to Lyell's rhetorical device and believe that gradual change is preferable (or even required) *a priori* because different meanings of uniformity are necessary postulates of method. Again and again in the history of geology after Lyell, we note reasonable hypotheses of catastrophic change, rejected out of hand by a false logic that brands them unscientific in principle. Thus, J Harlen Bretz's correct hypothesis for the formation of Washington's channeled scablands by catastrophic flooding was long dismissed by uniformitarians, who sought more time and many smaller rivers on little basis beyond a stated repugnance for catastrophes (several detractors at the famous 1927 confrontation between Bretz and scientists of the U.S. Geological Survey admitted that they had never visited the area, but were quite willing to propose gradualist alternatives as preferable *a priori*—see Baker and Nummedal, 1978; Gould, 1980). And the *New York Times,* in its editorial pages no less, has proclaimed that extraterrestrial impact as a catastrophic cause of the Cretaceous-Tertiary extinction has no place in science: "Terrestrial events, like volcanic activity or change in climate or sea level, are the most immediate possible cause of mass extinctions. Astronomers should leave to astrologers the task of seeking the causes of earthly events in the stars" (April 2, 1985).

Yet the Alvarez hypothesis of asteroidal or cometary impact is a powerful and plausible idea rooted in unexpected evidence of a worldwide iridium layer at the Cretaceous-Tertiary boundary, not developed from an anti-Lyellian armchair. It must be tested in the field, not dismissed *a priori*. In this light, and as a final example of how Lyell's rhetorical confusion might stifle legitimate research, I note Lyell's harsh dismissal of the seventeenth-century scientist William Whiston, because he dared to promote comets, and not earthly agents alone, as sources of geological change. Comets, I note, are now a favored mechanism for mass extinction under the Alvarez hypothesis: "He [Whiston] retarded the progress of truth, diverting men from the investigation of the laws of sublunary nature, and inducing them to waste time in speculations on the power of comets to drag the waters of the ocean over the land—on the condensation of the vapors of their tails into water, and other matters equally edifying" (I, 39).

Most geologists, especially if they believe the textbook cardboard they read as students, think that Lyell was the founder of modern practice in our profession. I do not deny that *Principles of Geology* was the most important, the most influential, and surely the most beautifully crafted work of nineteenth-century geology. Yet if we ask how Lyell's controlling vision has influenced modern geology, we must admit that current views represent a pretty evenly shuffled deck between attitudes held by Lyell and the catastrophists. We do adhere to Lyell's two methodological uniformities as a foundation of proper scientific practice, and we continue to praise Lyell for his ingenious and forceful defense. But uniformities of law and process were a common property of Lyell *and* his catastrophist opponents—and our current allegiance does not mark Lyell's particular triumph.

As for substantive uniformities of rate and state, our complex and multifarious world says yes and no to bits of both. Lyell himself abandoned uniformity of state for life's history, while a primary thrust of modern research into Precambrian strata (the first five-sixths of our earth's history!) tries to identify how the early earth differed—in sedimentary consequences of an atmosphere devoid of oxygen, for example—from the current order of nature. The great geologist Paul Krynine once called "uniformitarianism" (but meaning only uniformity of state) "a dangerous doctrine" because it led us to deny or underplay these early differences (Krynine, 1956). Even uniformity of rate, Lyell's stronger and more persistent argument, has suffered increasing attack as a generality. In the history of life, for example, alternative punctuational styles of change have been advocated at all levels—from the origin of species (punctuated equilibrium) to the overturn of entire faunas (catastrophic hypotheses of mass extinction).

Lyell, by the power of his intellect and the strength of his vision, deserves his status as the greatest of all geologists. But our modern understanding is not his, either unvarnished or even predominantly, but rather an inextricable and even mixture of uniformitarianism and catastrophism. Lyell won a rhetorical war, and cast his opponents into a limbo of antiscience, but we have been compelled to balance his dichotomy—because time's arrow and time's cycle both capture important aspects of reality.

Epilogue

Most working scientists are notorious for their lack of interest in history. In many fields, journals more than a decade old are removed from library shelves and relegated to microfiche, to unheated attics, or even to the junkpile.

During the summer of 1972, I met in Woods Hole with three of the best young-Turk paleontologists and ecologists of our day— Dave Raup, Tom Schopf, and Dan Simberloff. We were trying, immodestly to be sure and with limited success as it happened, to find a new approach to the study of life's history. We wanted to break away from a paleontological tradition that we found stultifying—a training that made professionals into experts about particular groups at particular times in particular places, and seemed to discourage any development of general theories that might be expressed in testable and quantitative terms. We decided to work with random models of origin and extinction, treating species as particles with no special properties linked to their taxonomic status or time of flourishing. As we proceeded, we realized that our models bore remarkable similarity in concept to Lyell's method for dating the Tertiary. Indeed, we recognized that his vision of time's stately cycle had become the ground of our proposal. And so, for several hours, four young scientists out to change the world sat around a table and talked about Charles Lyell.

My colleague Ed Lurie, distinguished scholar of Louis Agassiz, once told me that he had tried to escape Agassiz for years, and to branch into other areas of nineteenth-century American biology. But he couldn't, for Agassiz loomed so large that his shadow extended everywhere. Any exploration of any subfield in American biology became, at least in part, a study of Agassiz's influence.

I feel much the same way about Charles Lyell. I have made no active effort to avoid him, but neither do I court his presence. Still, I cannot escape him. If I recognize a baleful influence of his rhetoric, my quest for a different formulation still embraces another aspect of his vision. Thus, when Eldredge and I developed the theory of punctuated equilibrium, we tried, above all, to counteract both Lyell's bias of gradualism and his method of probing behind appearance to defend the uniformity of rate against evidence read literally—for punctuated equilibrium, as its essential statement, accepts the literal record of geologically abrupt appearance and subsequent stasis as a reality for most species, not an expression of true gradualism filtered through an imperfect fossil record. We felt mighty proud of ourselves for breaking what we saw as a conceptual lock placed by Lyell's vision upon the science of paleontology. But then, from another point of view, what is punctuated equilibrium but a nongradualist view of evolutionary theory applied to Lyell's *original* vision of species as discrete particles, arising at geological moments in space and time, and persisting unchanged until their extinction? Lyell had compromised this vision by embracing a gradualist account of evolution to salvage his uniformities, but we had been driven back to his original formulation. We had, it seemed, attacked Lyell in order to find him.

I could drown you in words—indeed I already have—about the power and importance of Lyell's vision. But any scientist will tell you that utility in practice is the only meaningful criterion of success. I can offer no greater homage to Charles Lyell than my personal testimony that he doth bestride my world of work like a colossus.

Figure 5.1
The entire composition of James Hampton's
Throne of the Third Heaven of the Nations' Millennium General Assembly,
showing its bilateral symmetry.

Boundaries

Hampton's Throne and Burnet's Frontispiece

In his play *The Road to Mecca*, Athol Fugard tells the true story of Helen Martins, an elderly Afrikaner, widow of a farmer, and inhabitant of the isolated Karroo village of New Bethesda. Late in life, she was struck with a vision, and began to construct elaborate statues of concrete, covered with glitter, in her backyard. These ecumenical monuments are religious in character, though not especially Christian, and they mark a road to her personal Mecca. Marius Byleveld, the local dominee (reverend), who secretly loves Helen but must preserve community standards, wants to remove her to a nursing home, more for reasons of kindness and genuine fear for her sanity than for simple reactionary motives of the Dutch Reformed Church. In a stirring scene, Helen confronts him: How can anyone claim that she is mad? A madman has lost contact with reality. But she, to build her beloved statues, had to learn to mix cement, and to grind beer cans in a coffee mill to produce bits of glitter.

James Hampton (1909–1964), a black man from Elloree, South Carolina, worked the evening shift as a janitor in various public buildings of Washington, D.C. Beginning in 1931, and recurring often thereafter, God and his angels visited Hampton in physical form and instructed him to build the throne room for Christ's second coming. In 1950, Hampton rented an unheated, poorly lit garage in a deteriorating neighborhood, telling his landlord that he was "working on something" that couldn't fit into his boarding-house room. There, until his death in 1964, he built one of the great works of American folk sculpture: *The Throne of the Third Heaven of the Nations' Millennium General Assembly,* now on display in the National Museum of American Art, in Washington, D.C.

Hampton would finish his janitorial duties at midnight and then work at the garage for five or six hours thereafter. Let no man judge with ill intent the motivation of James Hampton or Helen Martins. Their visions gave joy and purpose to lives that much of society might have deemed unworthy of notice, or in a final phase of decline.

Hampton's throne contains 177 separate pieces (see Figure 5.1), mounted on rising platforms, and symmetrically disposed about the central structure (Figure 5.2), presumably Christ's throne for his second coming. Hampton crafted his pieces with consummate ingenuity and patience, from bits and fragments of used or discarded objects. Most larger pieces are constructed upon a base of old furniture. The central throne is an armchair with faded red cloth cushions; two semicircular offertories are fashioned from a large round table, sawed in half. (Merchants in a used-furniture district near Hampton's garage recalled that he often browsed among their wares, and then returned with a child's wagon to haul away his treasures.) Other pieces have no such substantial base. Some are built up from layers of insulation board, others from hollow cardboard cylinders that had supported rolls of carpeting.

Around these foundations, Hampton wove, wrapped, nailed, and otherwise affixed his glittering ornaments. He scavenged the neighborhood for gold and aluminum foil—from store displays, cigarette boxes, kitchen rolls; he even paid neighborhood bums for the foil on their wine bottles, and carried a sack wherever he went to hold any bits and pieces found on the streets. He also gathered light-bulbs, desk blotters, sheets of plastic, insulation board, and kraft paper—all, apparently, from the trash bins of government buildings where he worked.

Figure 5.2
Christ's throne, the centerpiece of Hampton's composition.

Hampton used these materials to fashion elaborate ornaments. Foil wrapped around lightbulbs and jelly jars forms the main decoration of most structures, but he used kraft paper and cardboard as well to make wings and stars (also lined or covered with foil), and he built rows of knobs from balls of crumpled foil (or newspaper surrounded by foil), and he lined the edges of several tables with thin tubes of electrical cable covered by gold foil.

Although I marveled at these ingenious constructions when I first saw Hampton's *Throne* in Washington, I was even more stunned by the clear and intricate concept of the ensemble. The symmetries are overpowering—and entirely consistent. Each piece is bilaterally symmetrical about its central axis (Figures 5.2 and 5.3), like a human body (not "symmetrical along several axes," as Hartigan, 1976, wrote). This two-fold symmetry also permeates the entire design—for all 177 pieces show perfect bilateral symmetry around the central axis defined by Christ's throne (Figure 5.1). Each piece to the right of Christ's throne is matched by a corresponding object to the left, symmetrically wrought to the last intricate detail of cardboard star and foil-covered lightbulb.

I was already thinking about this book when I saw Hampton's throne, but did not know how or whether to proceed. The metaphor of time's arrow and time's cycle had unlocked, at least for me, the central meaning of three great and generally misinterpreted documents of my profession. I had worked out both the tension and resolution in Burnet's commingling of these metaphors in his wonderful frontispiece (see Figure 2.1). I was on a brief coffee break from a dull meeting and had wandered into the museum's vestibule, where Hampton's throne is displayed. I was attracted by the glitter, and walked over.

This book congealed during the next ten minutes, one of those magic moments in any scholar's life. I looked at Hampton's throne and I saw Burnet's frontispiece. The two structures are identical in concept; they display the same conflict and resolution between time's arrow of history and time's cycle of immanence. They are not merely similar in overall purpose; they are identical in intricate detail as well.

Figure 5.3
Another piece from the midline of Hampton's *Throne*. Note the
bilateral symmetry of each piece considered separately.

In Burnet's grand display of universal history, Christ stands atop a circle of globes, proclaiming (in Greek) his immanence in the famous passage from Revelation: I am Alpha and Omega. In Hampton's design, Christ's throne stands top and center, and Hampton used the same quotation to illustrate his plan—for his blackboard (Figure 5.4), with its sketch of his entire concept, proclaims at top center, I AM ALPHA AND OMEAG [*sic*] THE BEGINNING AND THE END.

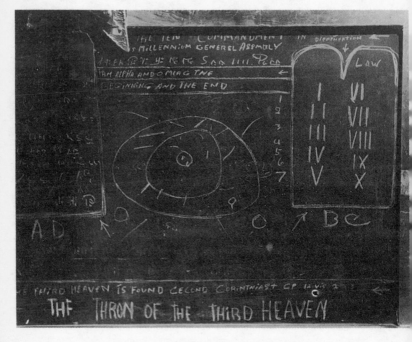

Figure 5.4

Hampton's blackboard, showing his plan for the entire composition. Note its near identity with Burnet's frontispiece: Christ stands at the top of each, proclaiming the same line from *Revelation*. History is a circle beginning and ending at the central Christ.

As Burnet's globes display the arrow of history beginning at Christ's left, reaching our present earth at the bottom, and chronicling our future at the right hand of Jesus, Hampton's pieces record the Old Testament, Moses, and the Commandments to the left side of Christ's throne, and the New Testament, Jesus, and Grace to the right. Two circular panels, mounted high in the top row to the left and right of Christ's throne, proclaim the status of each side—B.C. to the left, A.D. to the right (Figure 5.5). A row of plaques, found mounted on the left wall of Hampton's garage, bore names of the prophets, and a corresponding row on the right, the apostles.

But as the constructions of Burnet and Hampton display the arrow of history (in proper eschatological order with bad old pasts to the left, or sinister, side of divinity, and bright futures to the right), they also proclaim the immanence of God's glory. Not only does Christ directly announce his eternal presence in both, but (most important), the cycle of history follows an elaborate design with each incident on the old left precisely and symmetrically matching a corresponding replay on the new right.[1] In Burnet's frontispiece, we note a perfect world after consolidation of the elements from primeval chaos on the left, and our planet made perfect again after descent of the elements from the future universal fire in the same position on the right—or we find the earth destroyed by water

1. When I first saw Hampton's throne in Washington, this order was reversed in the display—with Old Testament objects to the right and New Testament to the left of Christ's throne. I wasn't bothered at first, reasoning that an uneducated man like Hampton might not have known the conventions of eschatology. Then I stopped myself: why should I make such an assumption. Who am I, a Jewish paleontologist on a coffee break, to presume superior knowledge of Christian tradition. So I wrote to the curator, asking if, perhaps, they had reversed the elements. They compared their arrangement with the photos of Hampton's order in his garage and wrote back to say that I was correct. Hampton had placed his pieces in conventional order, Old Testament objects to Christ's left. The display has now been returned to its original order. I don't say this to gloat (I guess I do, in part)—but it does speak for the power of time's cycle that a naturalist, knowing only Burnet's frontispiece, could recognize this misordering from knowledge of an abstract theme that pervades the centuries unchanged.

Figure 5.5
Plaques labeled B.C. and A.D., symmetrically placed about
the midline of Hampton's *Throne* to illustrate the direction
of time's arrow.

in Noah's flood to the left, and the earth consumed by fire to the right.

In Hampton's throne, each piece on the left (B.C.) matches a corresponding structure in the same position on the right (A.D.), built with the same symmetry to the last intricate detail—but proclaiming a different message. The structures are paired to unite Old Testament proclamations about the law and commandments with New Testament statements about salvation and grace. To cite just one example (Figure 5.6), a table on the left bears the following plaque (cited here in Hampton's arrangement and spelling):

> I am that I am
> Moses
> and God speak all
> these words saying
> I am the Lord thy God
> whih have brought
> the out of the land of
> Moses
> Commandment

The corresponding table on the right reads:

> St. James
> The second recorded of
> the ten commandment
> Recorded by St. James
> St. James
> Millennium

The themes of time's arrow and time's cycle, and their necessary union for any comprehensive view of our history and our estate, must be pervasive and powerful indeed if they could so motivate, in so essentially similar a way, the visions of both a great seventeenth-century scholar, private confessor to the King of England, and a poor black janitor in modern Washington, D.C.

The Deeper Themes of Arrows and Cycles

I have lectured for years in my introductory courses about the themes of time's arrow and time's cycle. Often, a student will ask, with that charming naiveté of a freshman who thinks that professors really do have simple answers to the deepest questions of the ages: Well, which is right? I always reply that the only possible answer can be "both and neither."

We often try to cram our complex world into the confines of what human reason can grasp, by collapsing the hyperspace of true conceptual complexity into a single line, and then labeling the ends of the line with names construed as polar opposites—so that all richness reduces to a single dimension and contrast of supposed opposites. All these dichotomies are false (or incomplete) because they can capture but a fraction of actual diversity, but one might be better (or at least more productive) than another because the limited axis of its particular contrast might express something more fundamental, more extensive in implication, or more in harmony with concerns of the actual debaters (see Chapter 1 for a fuller discussion of dichotomy).

I concluded that, if we must dichotomize, time's arrow and time's cycle is the most fruitful contrast for understanding the major issues underlying the greatest transformation that geology has or could contribute to human thought—the discovery of deep time. I reached this conclusion for several reasons: I believe that the major actors who struggled with time and the meaning of history from the late seventeenth through the mid-nineteenth century kept such a dichotomy at the forefront of their thought; thus, although arrow versus cycle may be as restrictively simplified as any single contrast, it was, at least, *their* dichotomy. For me, this contrast became a key for unlocking both the structure and meaning of great historical documents that I had read several times before, but had never understood or grasped as unified statements. But I also regard arrow versus cycle as a particularly "good" dichotomy because each of its poles captures a deep principle that human understanding of com-

Figure 5.6
Symmetrical structures from the right and left sides of Hampton's *Throne*,
showing in their inscription (see text) the repetitions of time's cycle.

plex historical phenomena requires absolutely—while other favored dichotomies, like evolution versus creation, cannot be so fruitful because the ends don't balance, at least in the sense that for certain classical issues (the history of life for example), one side is simply wrong, and therefore drops from intellectual interest, though not necessarily from political clout. Evolution consumed creation (in the strict version of all species fashioned *ex nihilo* on a young earth), but arrows and cycles need each other if we ever hope to grasp the meaning of history. Arrows and cycles are "eternal metaphors."

Time's arrow expresses the profundity in a style of explanation that many people find disappointing, or maximally unenlightening—the argument that "just history" underlies this or that phenomenon (not a law of nature, or some principle of timeless immanence). The essence of time's arrow lies in the irreversibility of history, and the unrepeatable uniqueness of each step in a sequence of events linked through time in physical connection—ancestral ape to modern human, sediments of an old ocean basin to rocks of a later continent. Abstracted parts of any totality may record the predictable (and repeatable) operation of nature's laws, but the details of an entire configuration are "just history" in the sense that they cannot arise again, and that another set of antecedents would have yielded a different outcome.

We may judge such a statement obvious and banal today, but it has often acted as the primary agent of major conceptual shift when history extended its realm, or when we learned to grasp history as contingent complexity, rather than preplanned harmony. For example, many popular pre-Darwinian taxonomic schemes were rooted in numerology—the grouping of all organisms into wheels of five, for example, with exact correspondences between spokes of all wheels—so that fishes on the wheel of vertebrates correspond with echinoderms on the wheel of all animals because both live exclusively in the sea, or mammals on the vertebrate circle with all vertebrates on the inclusive wheel, because both are the pinnacles of their respective systems. Such a scheme, proposed by William Swainson and other early nineteenth-century "quinarians" (see Figure 5.7), might work in an ahistorical world where organisms, like chemical elements on the periodic table, record timeless laws of nature, not complex contingencies of genealogy. Darwin removed the rationale for such numerologies in a single blow. The exterminating angel was history, not evolution itself. Some theories of evolution might permit such an ordered simplicity, but not Darwin's truly historical system with natural selection tracking a complex and unpredictable vector of climatic and geographic change, and with substantial randomness in the sources of variation. We tend to laugh at Swainson today, but his system was neither foolish nor irrational (and it was popular in his day). Quinarianism is intelligible in an ahistorical world, but we now know that taxonomic order is a product of "just history"—and the jumble of life's genealogies cannot fall into rigid circles of five.

Likewise, the ordered and predictable geological histories of Steno and Burnet are intelligible on a young earth imbued by a

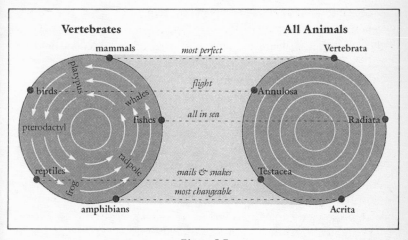

Figure 5.7
Swainson's rigidly numerological system of taxonomy, inconceivable for an arrangement of organisms in a world of contingent history.

watchful creator with signposts of his harmonious mind—but not on an ancient planet coursing for billions of years upon contingent pathways of "just history." The laws of plate tectonics may be simple and timeless, but they yield a complex uniqueness of results when we trace the actual configurations of continents through time.

Time's arrow of "just history" marks each moment of time with a distinctive brand. But we cannot, in our quest to understand history, be satisfied only with a mark to recognize each moment and a guide to order events in temporal sequence. Uniqueness is the essence of history, but we also crave some underlying generality, some principles of order transcending the distinction of moments—lest we be driven mad by Borges's vision of a new picture every two thousand pages in a book without end. We also need, in short, the immanence of time's cycle.

The metaphor of time's cycle captures those aspects of nature that are either stable or else cycle in simple repeating (or oscillating) series because they are direct products of nature's timeless laws, not the contingent moments of complex historical pathways. The geometry of space regulates how spheres of different sizes may fill a volume in arrangements of regular repetition—and the taxonomy of molecular order in minerals represents a compendium of these possibilities. In Linnaeus's time, many scientists hoped for a unified taxonomy of all natural objects, and named mineral "species" by the rules of binomial nomenclature developed for organisms. This effort has been abandoned as misguided by false perceptions of unity. Organisms follow time's arrow of contingent history; minerals, time's cycle of immanent geometrical logic. The dichotomous branching system of organic nomenclature captures the reality of historical diversification and genealogical connection; the order of minerals is differently established, and not well expressed by a system designed to classify items on a topology of continuous branching without subsequent coalescence.

But the order of minerals will bear meaningful comparison with other systems that reflect the same geometric laws, however disparate their objects. Several years ago, an international convention of mineralogists visited the Alhambra of Granada, the architectural primer for Islam's keen understanding of geometrical regularity in ornamentation. One of my colleagues noted with pleasure that patterns of symmetry in tile designs of the Alhambra included every two-dimensional arrangement recognized by mineralogists in earthly rocks.

This similarity of time's cycle teaches us something deep about nature's structure because the congruence of ions and tiles is *not* a product of "just history." The complex likenesses of organic genealogy are passive retentions from common ancestry—contingencies of historical pathways, not records of immanent regularities. (I type with the same bones used by a bat to fly, a cat to run, and a seal to swim because we all inherited our fingers from a common ancestor, not because laws of nature fashioned these bones independently, and in necessary arrangement.) The complex similarity of tile and mineral patterns records an active, separate development to the same result under immanent rules of natural order.

These two kinds of similarity—by genealogical connection, or time's arrow, and by separate reflection of the same immanent laws, or time's cycle—join forces when we try to unravel nature's complexity. The vision of time's cycle enabled Hutton and Lyell to grasp deep time, but we couldn't mark units within this immensity until time's arrow of the fossil record established a criterion of uniqueness for each moment. The neptunists failed to unlock stratigraphy because they falsely assumed that rocks carried signatures of temporal uniqueness. But rocks are simple objects, and their similarities designate formation under recurring conditions, not time's arrow of genealogy. The proper paleontological criterion, based we now know on contingent pathways of evolutionary change, allowed us to mark by time's arrow what the matrix of time's cycle had established.

Evolutionary biologists have long recognized, as *the* fundamental operation of our profession, the proper distinction between similarities of time's arrow and time's cycle. We designate them by different names—*homology* for passive retention of features shared by common ancestry along time's arrow of genealogy; *analogy* for active evolution of similar forms in separate lineages because immanent principles of function specify a limited range of solutions to common problems faced by organisms throughout time. The wings of birds, bats, and pterodactyls are analogous, though strikingly similar in aerodynamic design, because no common ancestor of any pair had wings—and flight evolved independently in three separate lineages. The detailed similarity in number and arrangement of arm-bones in humans, chimps, and baboons does not record a law of nature imposed upon separate productions, but simple inheritance from a common ancestor.

Thus, all taxonomists will tell you that they must, above all, separate analogous from homologous similarity, discard the analogies, and base classifications upon homologies alone—for taxonomies record pathways of descent. But any functional morphologist will pass over homologies as simple repetitions of the same experiment, and seek analogies that teach us about the limits of variety when separate lineages evolve structures for similar function.

The arrow of homology and the cycle of analogy are not warring concepts, fighting for hegemony within an organism. They interact in tension to build the distinctions and likenesses of each creature. They interweave and hold one another, as the laws of time's cycle mold the changing substances of history. The relentless arrow of history assures us that even the strongest analogy will betray signs of uniqueness, and permit a proper placement in taxonomy and time. *Ichthyosaurus* (Figure 5.8), descendant of terrestrial reptiles that returned to the sea, evolved the most uncanny resemblances to fishes—even developing, from no known precursor, a dorsal fin in the proper hydrodynamic position, and a tail fin with two symmetrical lobes, as principles of optimal swimming dictate. But ichthyosaurs retain signs of their reptilian heritage. The dorsal fin contains no bony supports as in fishes; the vertebral column bends into the lower lobe of the tail, not into the upper, or ceasing in the midline, as in fishes; the supports of the flippers are finger bones, not fin rays. In other words, the immanent and predictable features of good design are crafted from materials that preserve the stamp of time's arrow. "Fishness" is a timeless principle of good design; *Ichthyosaurus* is a particular reptile in a particular time and place. Two world views, eternal metaphors, jockey for recognition within every organism—receiving special attention according to the aims and interests of students: homology and analogy; history and optimality; transformation and immanence.

How, then, can we judge the interaction of time's arrow and time's cycle within each object? I can specify two incorrect approaches: we must not seek one in order to exclude the other (as Hutton did in denying history, before Lyell expanded his vision to become a historian of time's cycle); but neither should we espouse a form of wishy-washy pluralism that melds the end-members into an undefined middle and loses the essence of each vision—the uniqueness of history, and the immanence of law. Arrows and cycles, after all, are only categories of our invention, devised for

Figure 5.8
Time's arrow of homology and time's cycle of analogy combine to produce this *Ichthyosaurus*. The dorsal and caudal fins are convergent (with uncanny precision) upon similar structures in fishes, but have evolved independently in this descendant of terrestrial reptiles. Yet the signs of reptilian origin (homology) are evident throughout the skeleton, particularly in the finger bones enclosed within the fins.

clarity of insight. They do not blend, but dwell together in tension and fruitful interaction.

When Ritta-Christina (Figure 5.9), the Siamese twin girls of Sardinia, died in 1829, the pundits of her day debated at great length and to no avail the pressing issue of whether she had been one or two people. The issue could not be resolved because it has no answer expressed in terms these pundits sought. Their categories were wrong or limited. The boundaries between oneness and twoness are human impositions, not nature's taxonomy. Ritta-Christina, formed from a single egg that failed to divide completely in twinning, born with two heads and two brains but only one lower half, was in part one, and in part two—not a blend, not one-and-a-half, but an object embodying the essential definitions of both oneness and twoness, depending upon the question asked or the perspective assumed.

The same tension and multiplicity have pervaded our Western view of time. Something deep in our tradition requires, for intelligibility itself, both the arrow of historical uniqueness and the cycle of timeless immanence—and nature says yes to both. We see this tension in Burnet's frontispiece, in Lyell's method for dating the Tertiary, and in Hampton's *Throne*. We find it etched in pictorial form into the iconography of any medieval cathedral in Europe, where the arrow of progressive history passes from Old Testament lore on the dark north side, to resurrection and future bliss on the sunlit south. Yet we also see the cycle within the arrow. A set of correspondences—like Burnet's globes and Hampton's symmetries—teaches us that each event of Christ's life replays an incident in the previous cycle of Old Testament history. We see, in short, Burnet's resolution of rolling wheels. Each moment of the replay is similar as a reflection of timeless principles, and different because time's wheel has moved forward.

In the twelfth-century stained glass of Canterbury (Figure 5.10), Lot's wife turns into a pillar of salt, her whiteness contrasting starkly with the glittering colors of Sodom and Gomorrah in flames. In the corresponding panel of the second cycle, angels visit the wise

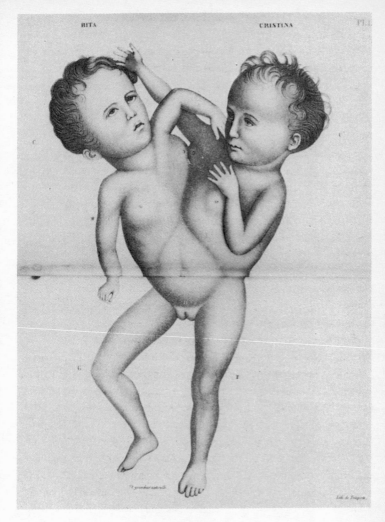

Figure 5.9
Ritta-Christina, the Siamese twins of Sardinia—neither two nor one person
but residing at an undefined middle of this continuum.

Figure 5.10
Time's cycle in Canterbury. The tale of Lot's wife is repeated
in the angel's advice to the Magi: do not return to Herod.

Figure 5.11
Ceiling bosses of Norwich Cathedral. Noah in the ark corresponds
with the baptism of Jesus.

Figure 5.12
Painted and stained glass windows of King's College, Cambridge.
Jonah emerging from the belly of the great fish corresponds with Christ
rising from the tomb.

Figure 5.13
From the great south window of Chartres. Time's arrow and cycle
connect as the gospel writers of the New Testament are shown as
dwarfs seated upon the shoulders of Old Testament prophets.

Figure 5.14
From Chartres Cathedral. At the end of time,
the just rise to their beginning and reside
in the bosom of Abraham.

men in a dream and tell them to travel straightaway to their own country, and not to return to Herod. The common message: don't look back.

On the ceiling bosses of Norwich (Figure 5.11), Noah and the animals ride out the deluge in an ark, while John baptizes Jesus with waters of salvation in the second cycle.

In the sixteenth-century painted glass of King's College Chapel in Cambridge (Figure 5.12), Jonah, expelled from the belly of the great fish, represents Christ resurrected from the tomb—because both faced the darkness of death and rose again on the third day.

But the great south window of Chartres presents the finest illustration in all our art of the necessary interaction between arrows and cycles for any comprehensive view of history (Figure 5.13). Here, at the close of the second cycle, the gospel writers, scribes of the New Testament, appear as dwarfs sitting on the shoulders of Isaiah, Jeremiah, Ezekiel, and Daniel, the great prophets of time's first cycle. To see farther, as Newton remarked to crown four centuries of metaphor dating back to these windows (Merton, 1965), we must stand on the shoulders of giants.

If we step outside and study the statues on the porch of Chartres (Figure 5.14), we discover the epitome of this book in a single figure. Burnet's history has run its course; the thousand-year reign of Christ on earth is over. The righteous have ascended to eternal reward. But they do not go forward into further uniqueness. They rise instead to the beginning and come to rest in the bosom of Abraham, the patriarch. James Hampton would have understood, for his vision embraced both metaphors of time. Rock a my soul . . .

BIBLIOGRAPHY

INDEX

Bibliography

Aristotle. 1960 ed. *Organon (Posterior Analytics)*. Trans. H. Fredennick. Loeb Classical Library, no. 39. Cambridge, Mass.: Harvard University Press.

Baker, V. R., and D. Nummedal. 1978. *The Channeled Scabland*. Washington, D.C.: National Aeronautics and Space Administration, Planetary Geology Program.

Borges, J. L. 1977. *The Book of Sand*. Trans. N. T. di Giovanni. New York: Dutton.

Bradley, S. J. 1928. *The Earth and Its History*. Boston: Ginn & Company.

Buckland, F. 1857, 1874 ed. *Curiosities of Natural History*. London: Richard Bentley and Son.

Buckland, W. 1836, 1841 ed. *Geology and Mineralogy Considered with Reference to Natural Theology*. Philadelphia: Lea & Blanchard.

Burnet, T. 1680–1689. *Telluris theoria sacra*. London.

———— 1691. *Sacred Theory of the Earth*. London: R. Norton.

———— 1965. *The Sacred Theory of the Earth*. With an introduction by B. Willey. Carbondale: Southern Illinois University Press.

Butterfield, H. 1931. *The Whig Interpretation of History*. London: G. Bell.

Chorley, R. J., A. J. Dunn, and R. P. Beckinsale. 1964. *The History of the Study of Landforms*. Vol. 1: *Geomorphology before Davis*. London: Methuen.

CRM Books. 1973. *Geology Today*. Del Mar, Calif.

Darwin, C. 1859. *On the Origin of Species by Means of Natural Selection*. London: John Murray.

Davies, G. L. 1969. *The Earth in Decay: A History of British Geomorphology 1578–1878*. New York: American Elsevier.

D'Orbigny, A. 1849–1852. *Cours élémentaire de paléontologie et de géologie stratigraphique*. Paris.

Eiseley, L. 1959. *Charles Lyell*. Scientific American Reprint. San Francisco: W. H. Freeman.

Eldredge, N., and S. J. Gould. 1972. Punctuated equilibria: An alternative to phyletic gradualism. In Schopf, T. J. M., ed., *Models in Paleobiology,* pp. 82–115. San Francisco: Freeman, Cooper & Co.

Eliade, M., 1954. *The Myth of the Eternal Return*. Princeton: Princeton University Press.

Fenton, C. L., and M. A. Fenton. 1952. *Giants of Geology*. Garden City, N.Y.: Doubleday.

Geikie, A. 1905. *The Founders of Geology*. New York: Macmillan.

Gilluly, J., A. C. Waters, and A. O. Woodford. 1959. *Principles of Geology*. 2nd ed. 1968. San Francisco: W. H. Freeman.

Goodman, N. 1967. Uniformity and simplicity. *Geological Society of America Special Papers* 89: 93–99.

Gould, S. J. 1965. Is uniformitarianism necessary? *American Journal of Science* 263: 223–228.

——— 1970. Private thoughts of Lyell on progression and evolution. *Science* 169: 663–664.

——— 1979. Agassiz's marginalia in Lyell's *Principles,* or the perils of uniformity and the ambiguity of heroes. In W. Coleman and C. Limoges, eds., *Studies in the History of Biology,* pp. 119–138. Baltimore: The Johns Hopkins University Press. Festschrift dedicated to Ernst Mayr on his 75th birthday.

——— 1980. *The Panda's Thumb*. New York: W. W. Norton.

——— 1982. The importance of trifles (essay for the 100th anniversary of the death of Charles Darwin). *Natural History* 91, no. 4 (April).

——— 1983. *Hen's Teeth and Horse's Toes*. New York: W. W. Norton.

——— 1985. *The Flamingo's Smile*. New York: W. W. Norton.

——— 1986a. Evolution and the triumph of homology, or why history matters. *American Scientist* 74: 60–69.

——— 1986b. Archetype and ancestor. *Natural History* 95, no. 10 (October).

Greene, J. C. 1961. *The Death of Adam*. New York: Mentor Books.

Hartigan, L. R. 1977. The Throne of the Third Heaven of the Nations Millennium General Assembly. Montgomery, Ala.: Montgomery Museum of Fine Arts.

Hobbs, W. H. 1916. Foreword: The science of the Prodromus of Nicolaus Steno. In J. G. Winter, trans., The Prodromus of Nicolaus Steno's Dissertation, pp. 169–174. University of Michigan Studies, Humanistic Series, vol. 11. New York: Macmillan.

Hooykaas, R. 1963. *The Principle of Uniformity in Geology, Biology, and Theology.* Leiden: E. J. Brill.

Hutton, J. 1788. *Theory of the Earth. Transactions of the Royal Society of Edinburgh* 1: 209–305.

—— 1795. *Theory of the Earth with Proofs and Illustrations.* Edinburgh: William Creech.

Krynine, P. D. 1956. Uniformitarianism is a dangerous doctrine. *Journal of Paleontology* 30: 1003–1004.

Leet, L. D., and S. Judson. 1971. *Physical Geology.* 4th ed. Englewood Cliffs, N.J.: Prentice-Hall.

Longwell, C. R., R. Foster Flint, and J. E. Sanders. 1969. *Physical Geology.* New York: John Wiley.

Lyell, C. 1830–1833. *Principles of Geology, Being an Attempt to Explain the Former Changes of the Earth's Surface by Reference to Causes Now in Operation.* London: John Murray.

—— 1851. Anniversary address of the President. *Quarterly Journal of the Geological Society of London (Proceedings of the Geological Society)* 7:xxv–lxxvi.

—— 1862. *On the Geological Evidences of the Antiquity of Man.* London.

—— 1872. *Principles of Geology or the Modern Changes of the Earth and Its Inhabitants Considered as Illustrative of Geology.* 11th ed., 2 vols. New York: D. Appleton.

Lyell, K. M. 1881. *Life, Letters and Journals of Sir Charles Lyell.* 2 vols. London: John Murray.

Marvin, U. 1973. *Continental Drift.* Washington, D.C.: Smithsonian Institution Press.

Mayr, E. 1963. *Animal Species and Evolution.* Cambridge, Mass.: Harvard University Press.

McPhee, J. 1980. *Basin and Range.* New York: Farrar, Straus, and Giroux.

Merton, R. K. 1965. *On the Shoulders of Giants.* New York: Harcourt, Brace and World.

Mill, J. S. 1881. *A System of Logic.* 8th ed. Book 3, chap. 3, Of the ground of induction. London.

Morris, R. 1984. *Time's Arrows.* New York: Simon and Schuster.

Peirce, C. S., 1932. *Collected Papers.* Vol. 2: *Elements of Logic.* Ed. C. Hartshorne and P. Weiss. Cambridge, Mass.: Harvard University Press.

Playfair, J. 1802. *Illustrations of the Huttonian Theory of the Earth.* Edinburgh: William Creech.

——— 1805. Biographical account of the late Dr. James Hutton. In V. A. Eyles and G. W. White, eds., *James Hutton's System of the Earth . . . ,* pp. 143–203. New York: Hafner Press, 1973.

Porter, R. 1976. Charles Lyell and the principles of the history of geology. *British Journal for the History of Science* 9: 91–103.

Price, G. M. 1923. *The New Geology.* 2nd ed. Mountain View, Calif.: Pacific Press.

Rensberger, B. 1986. *How the World Works.* New York: William Morrow.

Rossi, P. 1984. *The Dark Abyss of Time.* Chicago: University of Chicago Press.

Rudwick, M. J. S. 1972. *The Meaning of Fossils.* London: Macdonald.

——— 1975. Caricature as a source for the history of science: De la Beche's anti-Lyellian sketches of 1831. *Isis* 66: 534–560.

——— 1976. The emergence of a visual language for geological science, 1760–1840. *History of Science* 14: 149–195.

——— 1978. Charles Lyell's dream of a statistical palaeontology. *Palaeontology* 21: 225–244.

——— 1985. *The Great Devonian Controversy.* Chicago: University of Chicago Press.

Schweber, S. S. 1977. The origin of the *Origin* revisited. *Journal of the History of Biology* 10: 229–316.

Seyfert, C. K., and L. A. Sirkin. 1973. *Earth History and Plate Tectonics.* New York: Harper and Row.

Spencer, E. W. 1965. *Geology/A Survey of Earth Science.* New York: Thomas Y. Crowell.

Steno, N. 1669. *De solido intra solidum naturaliter contento dissertationis prodromus.* Florence.

——— 1916. *The Prodromus of Nicolaus Steno's Dissertation.* Trans. J. G. Winter. University of Michigan Studies, Humanistic Series, vol. 11. New York: Macmillan.

Stokes, W. L. 1973. *Essentials of Earth History*. Englewood Cliffs, N.J.: Prentice-Hall.

Turnbull, H. W., ed. 1960. *The Correspondence of Isaac Newton*. Vol. 2: *1676–1687,* Cambridge: Cambridge University Press.

Whewell, W. (published anonymously). 1832. *Principles of Geology . . . by Charles Lyell, Esq., F.R.S., Professor of Geology in Kings College London*. Vol. 2. London. *Quarterly Review* 47: 103–132.

White, A. D. 1896. *A History of the Warfare of Science with Theology in Christendom*. 2 vols. New York: D. Appleton.

Wilson, L. G., ed. 1970. *Sir Charles Lyell's Scientific Journals on the Species Question*. New Haven: Yale University Press.

Winter, J. G. 1916. Introduction. In *The Prodromus of Nicolaus Steno's Dissertation,* trans. J. G. Winter, pp. 175–203. University of Michigan Studies, Humanistic Series, vol. 11. New York: Macmillan.

Woodward, H. B. 1911. *History of Geology*. London.

Index

Abduction, 7
Actualism, 120, 127, 128, 151, 173
Agassiz, Louis, 115–117, 126, 128, 129, 179
Age of mammals, 155–156
Alhambra of Granada, 197
Alvarez hypothesis, 176,. 177
Analogy, 198
Aristotle, 42, 49, 74, 75

Basin and Range (McPhee), 3, 61, 68–69
Bible, 5, 6, 11–12; Burnet and, 5, 23–40 passim, 50, 51, 58, 127; Steno and, 52, 53–54, 58; Lyell and, 112–113. *See also* Genesis
Blackwood's Magazine, 73
Bodies, human, 82–83, 141, 142
"Book of Sand" (Borges), 48–49
Borges, Jorge Luis: Borges's dilemma, 48–49, 56, 92, 196
Bradley, S. J., 68
Bretz, J Harlen, 176
Buckland, Frank, 98, 100–101, 141
Buckland, William, 98, 99–100, 101, 129
Buffon, G., 94
Burnet, Thomas, 4–15 passim, 20–59, 84, 195–196, 208; and time's arrow, 15, 22, 27, 41, 42–46, 49, 58, 82, 184, 187, 192; and time's cycle, 15, 22, 41, 46–48, 49–51, 58, 80, 82, 184, 187, 192; context of, 17; Hutton

and, 23, 85–86; Lyell and, 23, 130; textbook myths on, 23–24, 26, 51, 67, 104, 127; and uniformity of law, 127. *See also Sacred Theory of the Earth*
Butterfield, Herbert, 4–5

Canterbury, 200, 202, 207
Carboniferous period, 145
Cardboard histories, *see* Textbooks
Catastrophism, 7, 13, 29, 112–117, 122–140 passim, 175–178
Causes, 74–78, 130–132; final, 75–78, 79, 95, 97; efficient, 76, 77, 78, 79; modern, 128, 146
Cenozoic Era, 86, 150
Chartres, 206, 208
Christian traditions, 10, 11. *See also* Bible
Clerk, John, of Eldin, 60, 70
Climate: Lyell on, 140–141, 144–146
Comets, 41, 176–177
Conflagration: Burnet and, 36–38, 50
Cornell, Ezra, 25
Creation: days of, 5, 38–40; vs. evolution, 9, 169, 172, 194
Creationism, "scientific," 23–24
Cretaceous-Tertiary transition, 175–176
CRM Books, 68
Crustal collapse, theory of, 31–32, 33, 34, 53, 57, 58
Curiosities of Natural History (Buckland), 98, 100–101

Cuvier, Georges, 18, 72, 152; proof of extinction by, 86; catastrophism of, 113, 114, 129, 133
Cycles, *see* Time's cycle

Darwin, Charles, 1, 8–9, 59, 91, 173, 195; natural selection theory of, 7, 84, 90, 143–144, 169–171, 195; and Lyell, 15, 135, 170, 171; on imperfection, 43, 84, 135; historical inference by, 90; on Buckland, 99; in *Origin of Species,* 105, 121, 143–144; and gradualism, 121, 135, 174
Davies, G. L., 16, 24, 51, 71–72, 73
Davy, Humphrey, 138
Deep time, 1–8, 90; Hutton and, 2, 5–6, 9–10, 15, 61–80 passim, 97, 197; Lyell and, 2, 6, 9–10, 15, 197; McPhee and, 2, 3, 68–69
De la Beche, Henry, 98, 100–102, 104, 141, 166n
Descartes, R., 58
Deshayes, G., 161, 167
Desmarest, N., 154
Dichotomy, 8–11, 14–16, 18–19, 193–194
Diogenes Laertius, 8
Directionalism, 82, 129, 131, 132, 145, 146, 149. *See also* Progressionism; Time's arrow; Vectors
d'Orbigny, Alcide, 127–128, 129

Earthquakes, 148
Ecclesiastes, 11–12
Edinburgh, 17
Efficient causes, 76, 77, 78, 79
Eiseley, Loren, 112–113
Eldredge, Niles, 2–3, 133n, 179
Elements of Geology (Lyell), 150
Eliade, Mircea, 12, 13, 14, 16
Elie de Beaumont, L., 129, 130, 131–132
Empiricism, 5, 23; Burnet and, 23, 27; catastrophism and, 122–123, 133–134. *See also* Fieldwork; Science
England, 17
Eocene period, 161, 162, 174, 175
Erosion, 63, 147; Burnet on, 44; Hutton on, 63, 65–66, 67, 77, 90

Eternity, 48–49, 56, 92
Europe, 139, 156–157
Evolution, 1, 9, 84, 90, 194, 197–198; Lyell and, 15, 146–147, 158, 166, 169–173, 174; punctuated equilibrium and, 179. *See also* Natural selection, theory of
Explication des textes, 16–17
Extinction, 86, 172, 175–176

Faunal transitions, 133. *See also* Geological record
Fenton, C. L., 24, 51
Fenton, M. A., 24, 51
Fieldwork, 152; Hutton and, 5–6, 66–73; Lyell and, 114, 115, 134, 167
Final cause, 75–78, 79, 95, 97
Flint, R. Foster, 113
Flood: Burnet and, 28n, 30–36, 38, 40, 44–45, 50, 58, 127; Steno and, 53–54, 58; Cuvier on, 114
Fossils, 86–88, 137–138, 152, 154, 158, 197; Hutton on, 86–88; Lyell and, 110, 138–140, 157, 158, 159–165, 173
Founders of Geology (Geikie), 23, 67–68
Fraser, J. T., 16
Freud, Sigmund, 1
Fugard, Athol, 181

Galilean revolution, 1–2
Geikie, Archibald, 23, 67–68, 69, 93, 103
Genesis, 112. *See also* Creation; Flood
Geoffroy Saint-Hilaire, E., 9, 17, 18
Geological record, 133–143. *See also* Fossils; Strata
Geological Society of London, 23, 66, 152, 168
Geological Survey, U.S., 176
Geology Today (CRM), 68
George III, 91
Giants of Geology (Fenton and Fenton), 24
Gillispie, C. C., 16, 151
Gilluly, J., 113
Glorious Revolution, 17
Goethe, J. W. von, 17, 18–19
Goodman, Nelson, 120

Gradualism, *see* Rate, uniformity of
Granite, 6, 67, 70–72

Halley, E., 41
Halley's Comet, 41
Hampton, James, 180, 181–192, 200, 208
Hanson, N. R., 6
Hebrew view, *see* Judeo-Christian traditions
Historical inference: Hutton and, 90–91
History, 59, 178; whiggish, 4–5, 7, 9, 10, 14, 17, 122; Burnet and, 44–46, 49, 57; Steno and, 56, 57; Hutton's denial of, 80–97, 151–152, 199; Lyell on, 150–167. *See also* Textbooks; Time's arrow
History of Rome (Niebuhr), 155
Homology, 197–198
Hooykaas, Reijer, 16, 118
Hume, David, 17, 64
Hutton, James, 4, 7, 14, 15, 60–97, 120; and deep time, 2, 5–6, 9–10, 15, 61–80 passim, 97, 197; context of, 17; and Burnet, 23, 85–86; Lyell and, 66–67, 93n, 97, 114, 128–129; textbook myths on, 66–70, 104, 114, 150–151; history denied by, 80–97, 151–152, 199
Huxley, T. H., 93

Ichthyosaurus, 99–101, 103–104, 166n, 198–199
Illustrations of the Huttonian Theory of the Earth (Playfair), 61, 94. *See also* Playfair, John
Imperfection, 43, 84–86, 135
Induction, 119–120
Infinity, 48–49, 56, 92

Jefferson, T., 91–92
Judeo-Christian traditions, 10, 11, 13. *See also* Bible
Judson, S., 68

Kant, I., 130
King's College Chapel, Cambridge, 204, 207–208
Kirwan, R., 93n

Krynine, Paul, 177
Kuhn, T. S., 6

Laboratory geologists, 69
Lamarck, J., 86, 166, 170n, 172
Laplace, P., 130
Law (natural): uniformity of, 119–127 passim, 169, 172, 173, 177
Leet, L. D., 68
Linnaeus, C., 196
Longwell, C. R., 113
Lurie, E., 179
Lyell, Charles, 4, 7, 13, 14, 98–179, 194; and deep time, 2, 6, 9–10, 15, 197; and time's cycle, 15, 103–104, 113, 115, 124, 129, 132–168, 173, 174, 175, 178, 199; Tertiary dating by, 15, 139, 155–167, 174–176, 178, 200; and evolution, 15, 146–147, 158, 166, 169–173, 174; and Burnet, 23, 130; and catastrophists, 29, 112–117, 122–140 passim, 175–178; and Hutton, 66–67, 93n, 97, 114, 128–129. See also *Principles of Geology*

Maastricht beds, 175
Machine, world, *see* World machine
Manual of Elementary Geology (Lyell), 150
Marvin, U., 68
Maupertuis, P.-L., 9
Mayr, Ernst, 144
McPhee, John, 2, 3, 61, 68–70
Mesozoic period, 139, 156
Milton, J., 36
Miocene period, 161, 163
Morris, Richard, 12, 16
Mummies, 109–110
Myth of the Eternal Return (Eliade), 12
Myths, *see* Textbooks

Narrative: Burnet and, 44–46, 49, 57; Steno and, 56, 57; Hutton and, 84–86, 89, 91, 97. *See also* History
Natural law, *see* Law, uniformity of
Natural selection, theory of, 7, 84, 90, 143–144, 169–171, 195
Neptunians, 157

Newton, Isaac: Newtonian science, 208; Lyell and, 2, 67, 97; Burnet and, 27, 38–41, 127; Hutton and, 67, 78–79, 81–82, 92, 94, 97, 151
New York Times, 176
Niebuhr, B. G., 155
Nietzsche, F., 16
Nonprogression, *see* State, uniformity of
North America, 139
Norwich Cathedral, 203, 207
Numerology, 59, 194–195

Oligocene period, 164
On the Geological Evidences of the Antiquity of Man (Lyell), 168
Origin of Species (Darwin), 105, 121, 143–144
Owen, Richard, 9

Paleozoic period, 138–139, 140, 155, 156, 168
Paleocene period, 161
Pan-selectionism, 84
Paradise Lost (Milton), 36
Paradox of the soil, Hutton's 76–77, 79, 80–97, 143
Peirce, C. S., 6–7, 119
Perfection, 32, 35, 43, 84–86, 166. *See also* Imperfection
Physics, 27–29, 30–41, 130–131
Planets, 82, 83
Plate tectonics, 68–69, 196
Plato, 16, 106
Playfair, John, 15, 61–62, 63, 80, 93–97; on deep time, 62, 70, 79; on world machine, 65, 95; historical narrative used by, 80, 96–97, 152
Pleistocene, 164
Pliocene period, newer and older, 61
Porter, R., 118
Precambrian strata, 177
Price, George McCready, 23–24
Principles of Geology (Lyell), 4, 104–106, 107n, 119–126, 142–150, 152–153, 165–168, 177; deep time in, 6; illustrations from, 71, 106, 121, 147, 148, 162, 163; and De la Beche illustration, 101–102, 141, 166n;

Agassiz's copy of, 115–116, 117; on evolution, 146–147, 166, 170
Process, uniformity of, 120, 123, 126–127, 169, 172–173, 177
Prodromus (Steno), 15, 51–54
Progressionism, 158, 164, 170n; Burnet and, 45, 50; Lyell and, 137–142, 150, 168–173. *See also* State, uniformity of; Time's arrow; Whiggish history
Punctuated equilibrium, theory of, 2–3, 133n, 179
Purposes, *see* Causes

Quinarians, 194–195

Rate, uniformity of (gradualism), 120–138 passim, 143, 146, 151, 160, 164, 172–179 passim
Raup, David, 178
Reason, human, 141–142
Religion: vs. science, 24–26, 27, 28–30. *See also* Judeo-Christian traditions
Restoration, *see* Uplift
Rhetoric, Lyell's, 9, 178; on uniformities, 107–111, 119, 122, 123, 124, 176; against catastrophism, 130–131, 133, 140
Ritta-Christina, 200, 201
Road to Mecca (Fugard), 181
Rossi, Paolo, 3, 4, 16
Royal Society of Edinburgh, 70, 91–92
Rudwick, Martin J. S., 16, 18, 102, 118, 119, 165

Sacred Theory of the Earth (Burnet), 4, 42–43, 44, 46–47; frontispiece to, 18, 20, 21–22, 26–27, 30, 41, 50–51, 184–187, 200
Schopf, T. J. M., 178
Science, 5, 6, 7; Burnet and, 23–27, 28–30; vs. religion, 24–26, 27, 28–30; Steno and, 51; Hutton and, 68; Lyell and, 153–154
"Scientific creationism," 23–24
Secondary strata, 156, 174, 175
Sedgwick, A., 129
Seyfert, C. K., 68
Simberloff, Daniel, 178

Simplicity: principle of, 120

Sirkin, L. A., 68

Smith, Adam, 7, 17, 64

Smith, William, 152

Soil: Hutton's paradox of, 76–77, 79, 80–97, 143. *See also* Erosion; Uplift

Spencer, E. W., 114

State, uniformity of (nonprogression), 123–126, 127, 129, 134–146 passim, 151, 164–177 passim

Steno, Nicolaus, 15, 51–59, 63, 195–196

Stokes, W. L., 114

Strata, 86–87, 133, 157; Hutton on, 62–63, 87, 88–90, 91, 96. *See also* Unconformities

Swainson, William, 194–195

Tertiary period, 15, 139, 155–167, 174–176, 178, 200

Textbooks (cardboard histories), 4–7, 14, 26; on Burnet, 23–24, 26–27, 51, 67, 104, 127; Lyell and, 23, 66–67, 102–103, 104–105, 118, 126, 130, 134, 143, 150–151, 177; on Steno, 51, 52; on Hutton, 66–70, 104, 114, 150–151; on d'Orbigny, 127

Theory of the Earth (Hutton), 4, 61–66 passim, 70, 72, 93n, 96

Thompson, D'Arcy W., 17

Throne of the Third Heaven of the Nations' Millennium General Assembly (Hampton), 180, 181–192, 200

Time, *see* Deep time; Infinity; Time's arrow; Time's cycle

Time's arrow, 10–16, 58–59, 184, 193–208; Burnet and, 15, 22, 27, 41, 42–46, 49, 58, 82, 184, 187, 192; Steno and, 52–55, 58; Hutton and, 63, 80, 82, 97, 151; Lyell and, 103–104, 113, 115, 132, 136–137, 145, 146, 149, 151; Hampton and, 184, 187, 188, 192

Time's cycle, 11–16, 58–59, 184, 193–208; Hutton and, 15, 63, 66, 67, 69–70, 73–82, 86, 87–88, 89, 92, 93, 94, 95, 97, 151 (*see also* World machine); Lyell and, 15, 103–104, 113, 115,

124, 129, 132–168, 173, 174, 175, 178, 199 (*see also* State, uniformity of); Burnet and, 15, 22, 41, 46–48, 49–51, 58, 80, 82, 184, 187, 192; Steno and, 55–57, 58; Hampton and, 184, 187, 190, 192

Transactions of the Royal Society of Edinburgh, 70

Transmutation, *see* Evolution

Twain, Mark, 2

Unconformities, 61–63, 70–72, 88–89; and discovery of deep time, 6, 61–62, 67; illustration of, 60, 61–62; Playfair and, 61–62, 63, 70, 96

Uniformitarianism, 90, 176, 177; Hutton and, 66–67, 90, 114, 128–129, 151; Lyell and, 66–67, 105–129, 132, 133–138, 151, 168–177 passim. *See also* Rate, uniformity of; State, uniformity of

Uplift: Hutton on, 62–63, 65–66, 67, 77–79, 91, 129, 151; Lyell on, 129, 137

U.S. Geological Survey, 176

Vectors: Burnet and, 44, 45, 49, 50, 57; Steno and, 56, 57; Lyell and, 132, 136, 140, 149. *See also* Progressionism; Time's arrow

Velikovsky, Immanuel, 27–28

Wallace, A. R., 170

Warfare of Science with Theology (White), 24–26

Waters, A. C., 113, 151

Watt, James, 17, 64

Werner, Abraham Gottlob, 69, 154, 157

Whewell, William, 105

Whiggish history, 4–5, 7, 9, 10, 14, 17, 122

Whig Interpretation of History (Butterfield), 4–5

Whiston, William, 176–177

White, Andrew Dickson, 24–26

William III, King of England, 26

Wilson, L. G., 170

Winter, J. G., 53

Woodford, A. O., 113, 151
Woodward, Horace B., 23
World machine, Hutton's, 65–74 passim, 87, 89, 91, 95, 96; and Newtonian science, 78–79, 97; perfection of, 85; and Borges's dilemma, 92; Lyell and, 97–98, 129

FOR THE BEST IN PAPERBACKS, LOOK FOR THE

In every corner of the world, on every subject under the sun, Penguin represents quality and variety – the very best in publishing today.

For complete information about books available from Penguin – including Pelicans, Puffins, Peregrines and Penguin Classics – and how to order them, write to us at the appropriate address below. Please note that for copyright reasons the selection of books varies from country to country.
